Collins

ROYAL
OBSERVATORY
GREENWICH

2015 GUIDE
to the
NIGHT SKY

Storm Dunlop and Wil Tirion

D1339437

Published by Collins
An imprint of HarperCollins Publishers
Westerhill Road
Bishopbriggs
Glasgow G64 2QT
www.harpercollins.co.uk

In association with
Royal Museums Greenwich, the group name for the National Maritime Museum,
Royal Observatory Greenwich, Queen's House and *Cutty Sark* 2014
www.rmg.co.uk

ISBN 978-0-00-759868-7

10 9 8 7 6 5 4 3 2 1

British Library Cataloguing in Publication Data
A catalogue record for this book is available from the British Library

Printed in China by South China Printing Co. Ltd

If you would like to comment on any aspect of this book, please contact us at the above address or online.
collinsmaps@harpercollins.co.uk
 facebook.com/collinsmaps
 @collinsmaps

Contents

Month-by-month guide

Introduction

The aim of this Guide is to help people find their way around the night sky, by showing how the stars that are visible change from month to month and by including details of various events that occur throughout the year. The objects and events described may be observed with the naked eye, or nothing more complicated than a pair of binoculars.

The conditions for observing naturally vary over the course of the year. During the summer, twilight may persist throughout the night and make it difficult to see the faintest stars. There are three recognized stages of twilight: civil twilight, when the Sun is less than 6° below the horizon; nautical twilight, when the Sun is between 6° and 12° below the horizon; and astronomical twilight when the Sun is between 12° and 18° below the horizon. Full darkness occurs only when the Sun is more than 18° below the horizon. During nautical twilight, only the very brightest stars are visible. During astronomical twilight, the faintest stars visible to the naked eye may be seen directly overhead, but are lost at lower altitudes. As the diagram shows, during the summer months full darkness never occurs at the latitude of London, and at Edinburgh

nautical twilight persists throughout the whole night, so at that latitude only the very brightest stars are visible.

Another factor that affects the visibility of objects is the amount of moonlight in the sky. At Full Moon, it may be very difficult to see some of the fainter stars and objects, and even when the Moon is at a smaller phase it may seriously interfere with visibility if it is near the stars or planets in which you are interested. A full lunar calendar is given for each month and may be used to see when nights are likely to be darkest and best for observation.

The celestial sphere

All the objects in the sky (including the Sun, Moon, and stars) appear to lie at some indeterminate distance on a large sphere, centred on the Earth. This *celestial sphere* has various reference points and features that are related to those of the Earth. If the Earth's rotational axis is extended, for example, it points to the North and South Celestial Poles, which are thus in line with the North and South Poles on Earth. Similarly, the *celestial equator* lies in the same plane as the Earth's equator, and divides the sky into northern and

The duration of twilight throughout the year at London and Edinburgh.

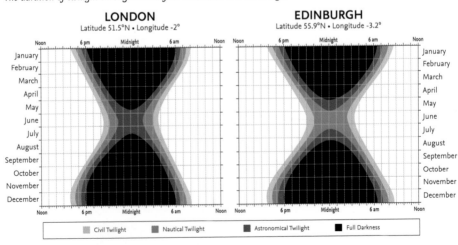

southern hemispheres. Because this Guide is written for use in Britain and Ireland, the area of the sky that it describes includes the whole of the northern celestial hemisphere and those portions of the southern that become visible at different times of the year. Stars in the far south, however, remain invisible throughout the year, and are not included.

It is useful to know some of the special terms for various parts of the sky. As seen by an observer, half of the celestial sphere is invisible, below the horizon. The point directly overhead is known as the **zenith**, and the (invisible) one below one's feet as the **nadir**. The line running from the north point on the horizon, up through the zenith and then down to the south point is the **meridian**. This is an important invisible line in the sky, because objects are highest in the sky, and thus easiest to see, when they cross the meridian in the south. Objects are said to **transit**, when they cross this line in the sky.

In this book, reference is frequently made in the text and in the diagrams to the standard compass points around the horizon. The position of any object in the sky may be described by its **altitude** (measured in degrees above the horizon), and its **azimuth** (measured

in degrees from north 0°, through east 90°, south 180°, and west 270°). Experienced amateurs and professional astronomers also use another system of specifying locations on the celestial sphere, but that need not concern us here, where the simpler method will suffice.

The celestial sphere appears to rotate about an invisible axis, running between the North and South Celestial Poles. The location (i.e., the altitude) of the Celestial Poles depends entirely on the observer's position on Earth or, more specifically, their latitude. The charts in this book are produced for the latitude of 50°N, so the North Celestial Pole (NCP) is 50° above the northern horizon. The fact that the NCP is fixed relative to the horizon means that all the stars within 50° of the pole are always above the horizon and may, therefore, always be seen at night, regardless of the time of year. The northern circumpolar region is an ideal place to begin learning the sky, and ways to identify the circumpolar stars and constellations will be described shortly.

The ecliptic and the zodiac

Another important line on the celestial sphere is the Sun's apparent path against the background stars – in reality the result

Measuring altitude and azimuth on the celestial sphere.

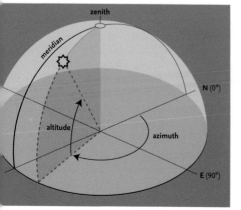

The altitude of the North Celestial Pole equals the observer's latitude.

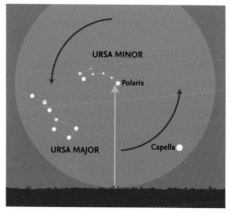

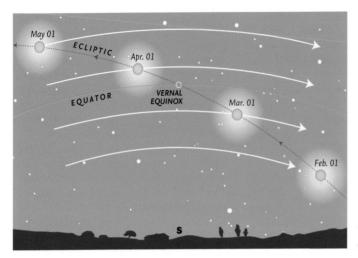

The Sun crossing the celestial equator in spring.

of the Earth's orbit around the Sun. This is known as the **ecliptic**. The point where the Sun, apparently moving along the ecliptic, crosses the celestial equator from south to north is known as the vernal (or spring) equinox, which occurs on March 20. At this time (and at the autumnal equinox, on September 22 or 23, when the Sun crosses the celestial equator from north to south) day and night are almost exactly equal in length. (There is a slight difference, but that need not concern us here.) The vernal equinox is currently located in the constellation of Pisces, and is important in astronomy because it defines the zero point for a system of celestial coordinates, which is, however, not used in this Guide.

The Moon and planets are to be found in a band of sky that extends 8° on either side of the ecliptic. This is because the orbits of the Moon and planets are inclined at various angles to the ecliptic (i.e., to the plane of the Earth's orbit). This band of sky is known as the Zodiac and, when originally devised, consisted of twelve **constellations**, all of which were considered to be exactly 30° wide. When the constellation boundaries were formally established by the International Astronomical Union in 1930, the exact extent of most constellations was altered and, nowadays, the ecliptic passes through thirteen constellations. Because of the boundary changes, the Moon and planets may actually pass through several other constellations that are adjacent to the original twelve.

The constellations

Since ancient times, the celestial sphere has been divided into various constellations, most dating back to antiquity and usually associated with certain myths or legendary people and animals. Nowadays, the boundaries of the constellations have been fixed by international agreement and their names (in Latin) are largely derived from Greek or Roman originals. Some of the names of the most prominent stars are of Greek or Roman origin, but many are derived from Arabic names. Many bright stars have no individual names and, for many years, stars were identified by terms such as 'the star in Hercules' right foot'. A more sensible scheme was introduced by the German astronomer Johannes Bayer in the early seventeenth century. Following his scheme – which is still used today – most of the brightest stars are identified by a Greek letter followed by the genitive form of the constellation's Latin name. An example is the Pole Star, also known as Polaris and α Ursae Minoris (abbreviated α UMi). The Greek alphabet is shown on page 93 with a list of all the constellations that may be seen from latitude 50°N, together with abbreviations, their genitive forms and English names. Other naming schemes exist for fainter stars, but are not used in this book.

Asterisms

Apart from the constellations (88 of which cover the whole sky), certain groups of stars, which may form a part of a larger constellation or cross several constellations, are readily recognizable and have been given individual names. These groups are known as *asterisms*, and the most famous (and well-known) is the 'Plough', the common name for the seven brightest stars in the constellation of Ursa Major, the Great Bear. The names and details of some asterisms mentioned in this book are given in the list on page 94.

Magnitudes

The brightness of a star, planet or other body is frequently given in magnitudes (mag.). This is a mathematically defined scale where larger numbers indicate a fainter object. The scale extends beyond the zero point to negative numbers for very bright objects. (Sirius, the brightest star in the sky is mag. -1.4.) Most observers are able to see stars down to about mag. 6, under very clear skies.

The Moon

As it gradually passes across the sky from west to east in its orbit around the Earth, the Moon moves by approximately its diameter (about half a degree) in an hour. Normally, in its orbit around the Earth, the Moon passes above or below the direct line between Earth and Sun (at New Moon) or outside the area obscured by the Earth's shadow (at Full Moon).

Occasionally, however, the three bodies are more-or-less perfectly aligned to give an eclipse: a solar eclipse at New Moon or a lunar eclipse at Full Moon. Depending on the exact circumstances, a solar eclipse may be merely partial (when the Moon does not cover the whole of the Sun's disk); annular (when the Moon is too far from Earth in its orbit to appear large enough to hide the whole of the Sun); or total. Total and annular eclipses are visible from very restricted areas of the Earth, but partial eclipses are normally visible over a wider area.

Somewhat similarly, at a lunar eclipse, the Moon may pass through the outer zone of the Earth's shadow, the penumbra (in a penumbral eclipse, which is not generally perceptible to the naked eye), so that just part of the Moon is within the darkest part of the Earth's shadow, the umbra (in a partial eclipse); or completely within the umbra (in a total eclipse). Unlike solar eclipses, lunar eclipses are visible from large areas of the Earth.

Occasionally, as it moves across the sky, the Moon passes between the Earth and individual planets or distant stars giving rise to an *occultation*. As with solar eclipses, such occultations are visible from restricted areas of the world.

The Planets

Because the planets are always moving against the background stars, they are treated in some detail in the monthly pages and information is given when they are close to other planets, the Moon or any of five bright stars that lie near the ecliptic. Such events are known as *appulses* or, more frequently, as *conjunctions*. (There are technical differences in the way these terms are defined – and should be used – in astronomy, but these need not concern us here.) The positions of the planets are shown for every month on a special chart of the ecliptic.

The term conjunction is also used when a planet is either directly behind or in front of

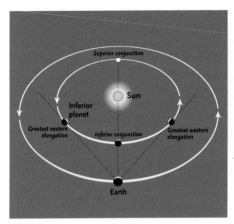

Inferior planet.

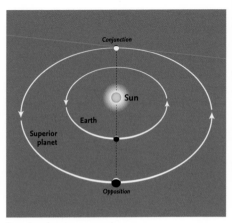

Superior planet.

the Sun, as seen from Earth. (Under normal circumstances it will then be invisible.) The conditions of most favourable visibility depend on whether the planet is one of the two known as *inferior planets* (Mercury and Venus) or one of the three *superior planets* (Mars, Jupiter and Saturn) that are covered in detail. (Certain details of the fainter superior planets, Uranus and Neptune, are included in this Guide, and special charts for both are given on page 18.)

The inferior planets are most readily seen at eastern or western *elongation*, when their angular distance from the Sun is greatest. For superior planets, they are best seen at *opposition*, when they are directly opposite the Sun in the sky, and cross the meridian at local midnight.

It is often useful to be able to estimate angles on the sky, and approximate values may be obtained by holding one hand at arm's length. The various angles are shown in the diagram, together with the separations of the various stars in the Plough.

Meteors

At some time or other, nearly everyone has seen a *meteor* – a 'shooting star' – as it flashed across the sky. The particles that cause meteors – known technically as 'meteoroids' – range in size from that of a grain of sand (or even smaller), to the size of a pea. On any night of the year there are occasional meteors, known as *sporadics*, that may travel in any direction. These occur at a rate that is normally between three and eight in an hour. Far more important, however, are *meteor showers*, which occur at fixed periods of the year, when the Earth encounters a trail of particles left behind by a comet or, very occasionally, by a minor planet (asteroid). Meteors always appear to diverge from a single point on the sky, known as the *radiant*, and the radiants of major showers are shown on the charts. Meteors that come from a circular area 8° in diameter around the radiant are classed as belonging to the particular shower. All others that do not come from that area are sporadics (or, occasionally from another shower that is active at the same time). A list of the major meteor showers is given on page 17.

Although the positions of the various shower radiants are shown on the charts, looking directly at the radiant is not the most effective way of seeing meteors. They are most likely to be noticed if one is looking about 40–45° away from the radiant position. (This is approximately two hand-spans as shown in the diagram for measuring angles.)

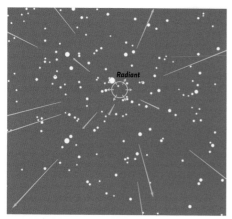

Meteor shower.

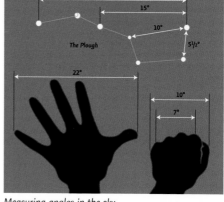

Measuring angles in the sky.

Other objects

Certain other objects may be seen with the naked eye under good conditions. Some were given names in antiquity – Praesepe is one example – but many are known by what are called 'Messier numbers', the numbers in a catalogue of nebulous objects compiled by Charles Messier in the late eighteenth century. Some, such as the Andromeda Galaxy, M31, and the Orion Nebula, M42, may be seen by the naked eye, but all those given in the list will benefit from the use of binoculars.

Apart from galaxies, such as M31, which contain thousands of millions of stars, there are also two types of cluster: open clusters, such as M45, the Pleiades, which may consist of a few dozen to some hundreds of stars; and globular clusters, such as M13 in Hercules, which are spherical concentrations of many thousands of stars. One or two gaseous nebulae, consisting of gas illuminated by stars within them, are also visible. The Orion Nebula, M42, is one, and is illuminated by the group of four stars, known as the Trapezium, which may be seen within it by using a good pair of binoculars.

Some interesting objects

Messier number	Name Constellation	Type Maps (months)
—	*Hyades* Taurus	open cluster Sep. – Apr.
—	*Double Cluster* Perseus	open cluster All year
M8	*Lagoon Nebula* Sagittarius	gaseous nebula Jun. – Sep.
M11	*Wild Duck Cluster* Scutum	open cluster May – Oct.
M13	*Hercules Cluster* Hercules	globular cluster Feb. – Nov.
M15	— Pegasus	globular cluster Jun. – Dec.
M22	— Sagittarius	globular cluster Jun. – Sep.
M27	*Dumbbell Nebula* Vulpecula	planetary nebula May – Dec.
M31	*Andromeda Galaxy* Andromeda	galaxy All year
M35	— Gemini	open cluster Oct. – May.
M42	*Orion Nebula* Orion	gaseous nebula Nov. – Mar.
M44	*Praesepe* Cancer	open cluster Nov. – Jun.
M45	*Pleiades* Taurus	open cluster Aug. – Apr.

The Northern Constellations

The northern circumpolar stars

In learning to find one's way around the sky and to identify the constellations, the northern circumpolar stars are the key starting point for northern-hemisphere observers. Not only are they visible at any time of the year, but nearly everyone living in the northern hemisphere is familiar with the seven stars of the Plough – known as the Big Dipper in North America – an asterism that forms part of the large constellation of Ursa Major. This is where we start. There are five main constellations and one all-important star to identify:

- *Ursa Major* (The Great Bear)
- *Polaris*
- *Ursa Minor* (The Little Bear)
- *Cassiopeia*
- *Cepheus*
- *Draco* (The Dragon)

Ursa Major

Because of the movement of the stars caused by the passage of the seasons, **Ursa Major** lies in different parts of the evening sky at different periods of the year. The diagram shows its position for the four main seasons to give you

> Continued on page 12

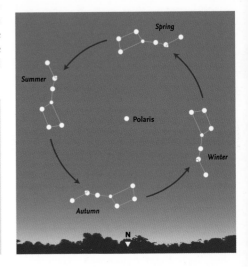

The seven stars of the Plough, the most prominent part of the constellation of Ursa Major, are easily recognized and visible throughout the year in the northern sky.

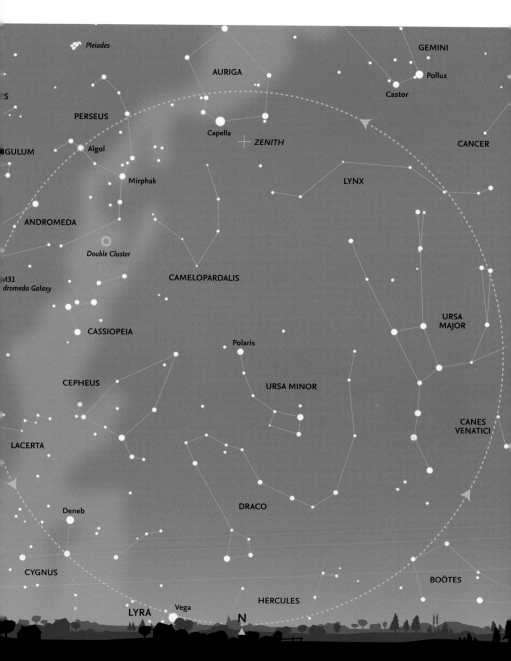

The stars and constellations inside the circle are always above the horizon, seen from our latitude.

a guide as to where to look. The seven stars remain visible throughout the year anywhere north of latitude 40°N. Even at the latitude (50°N) for which the charts in this book are drawn, many of the stars in the southern portion of the constellation are hidden below the horizon for part of the year.

Ursa Major is a very large constellation, but initially few people are familiar with the extended groups of stars that form part of it and lie well to the south and east. In the centre of the curve of stars that form the 'tail' of the Great Bear (or the 'handle' of the Dipper) lies the small constellation of Canes Venatici, which consists of two moderately bright stars and a scattering of fainter ones.

Polaris

Once you have identified the seven stars of the Plough, locate the stars, α and β Ursae

Majoris farthest away from the 'tail'. These two, named **Dubhe** and **Merak**, respectively, are known as the 'Pointers'. A line from Merak (the southernmost) to Dubhe, extended to about five times their separation, leads more-or-less directly to a fairly isolated bright star. This is the Pole Star, **Polaris**, or α Ursae Minoris and the brightest star in the constellation. All the stars in the northern sky appear to rotate around it. In fact, it lies slightly less than one degree away from the true pole, and a trailed photograph of the northern sky shows that it traces a tiny circle (about 1.5 degrees across) around the pole.

Apart from this small variation, the elevation of Polaris above the northern horizon is always equal to the observer's latitude. For sailors, this was an extremely useful property, and was used as a guide to navigation by generations of sailors from the time of the

Finding Polaris and Ursa Minor.

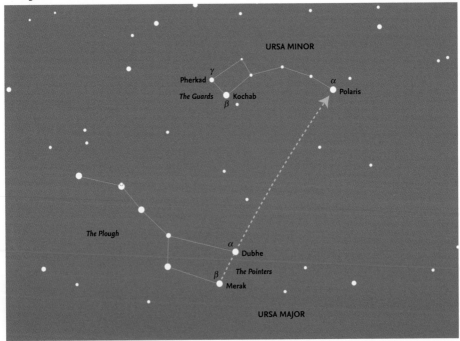

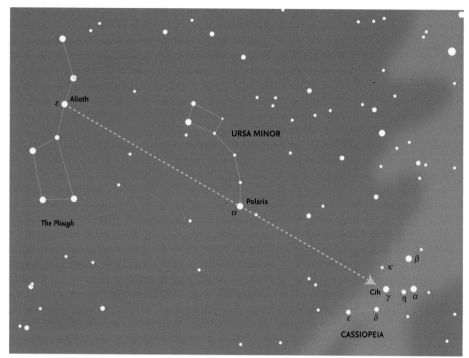

Finding Cassiopeia from the Plough and Polaris.

Greeks to comparatively modern times. Now it is our key to learning the sky.

Ursa Minor

In contrast to Ursa Major, **Ursa Minor** is a small constellation, consisting of little more than the relatively faint stars that form the 'Little Dipper' (as it is known in North America). Strangely, the asterism has no common name in Britain, other than the Little Bear, despite being superficially similar to the seven stars of the Plough. Just five stars – including a moderately close pair – form the body of the constellation, with another three forming the 'tail' at the tip of which lies Polaris – the end of the 'handle' of the Little Dipper. The two stars farthest from the pole, **Kochab** and **Pherkad** (β and γ Ursae Minoris, respectively) are known in English-speaking countries as 'The Guards'.

Cassiopeia

On the opposite side of Polaris and the North Pole from Ursa Major lies **Cassiopeia**. It has a highly distinctive shape, appearing as five stars that form a letter 'W' or 'M' depending on its orientation. Provided the sky is reasonably clear of clouds, you will nearly always be able to see either Ursa Major or Cassiopeia, and thus be able to orientate yourself on the sky.

To find Cassiopeia from Ursa Major, start with **Alioth** (ε Ursae Majoris), the first star in the tail of the Bear. A line from this star extended through Polaris points directly towards **Cih** (γ Cassiopeiae), the central star of the five. Cassiopeia lies right in the centre of the band of the Milky Way, so on a clear night a large number of fainter stars surround the five main stars, and this may sometimes make identification slightly more difficult for

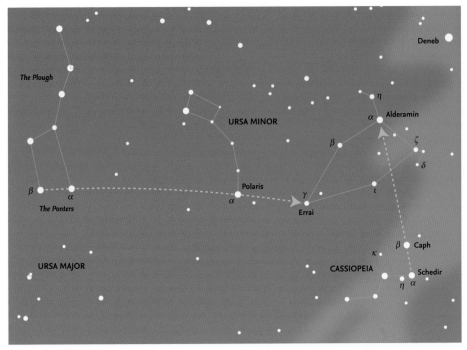

Locating the constellation of Cepheus.

beginners. Even when the Milky Way is not readily visible because of moonlight or light pollution, some people are still occasionally confused by the presence of the moderately bright stars, η and κ Cassiopeiae, that lie nearby. With just a little practice, however, Cassiopeia becomes easy to recognize.

Cepheus

Although the constellation of **Cepheus** is circumpolar, it is not nearly as well known as Ursa Major, Ursa Minor or Cassiopeia. This is partly because it does not form a highly distinctive pattern on the sky, and also because some of its stars are faint. Its shape is often likened to (and does actually resemble) the gable end of a house, with the 'ground' lying in the Milky Way not far from Cassiopeia, and the tip of the gable pointing towards the pole. Its brightest star, **Alderamin** (α Cephei), lies in

the Milky Way region, at the 'bottom right-hand corner' of the figure. A line from the stars **Schedir** and **Caph** (α and β Cassiopeiae), extended about three times their separation, passes just north of Alderamin. The star at the tip of the 'gable', **Errai** (γ Cephei) lies close to the line from Polaris (α UMi) to Caph (β Cas). The line from the Pointers to Polaris, if extended, also passes just 'below' the tip of the 'gable'.

Draco

The last of the circumpolar constellations that is fairly easy to identify is **Draco**, although it is such a long, winding constellation that it takes some experience to recognize it easily. It consists of a quadrilateral of stars, known, logically enough, as the 'Head of Draco' (and also the 'Lozenge'), and a long chain of stars forming the neck and body of the dragon.

Finding the Head of Draco is not particularly easy using just circumpolar stars. Locate the two stars Phad and Megrez (γ and δ Ursae Majoris) at the opposite end of the bowl of the Plough from the Pointers. Extend a line from Phad (γ UMa) through Megrez (δ UMa) by about eight times their separation. This takes you right across the sky below the Guards in Ursa Minor, and close to **Grumium** (ξ Draconis) at one corner of the quadrilateral. The brightest star in the constellation, **Etamin** (γ Draconis) lies farther to the south. From the head of Draco, the constellation first runs northeast to δ and ε Dra, then doubles back southwards, before winding its way round between Ursa Minor and Ursa Major, through **Thuban** (α Draconis), before ending at **Giausar**

(λ Draconis) almost on the line between the Pointers and Polaris.

The circumpolar chart shows that there is a large area of sky between Ursa Major and Cassiopeia that has relatively few bright stars. There is one constellation here, **Camelopardalis**, which is always circumpolar, but which is so faint that it is easier to learn to find it after you have gained some experience, and are familiar with some of the constellations (Perseus and Auriga) that lie farther south. Another dim constellation, **Lynx** – a long, straggling line of faint stars – is also rather difficult to see. Large parts of the constellations of Auriga, Cygnus, and Perseus are often circumpolar, depending on your exact latitude.

Finding the head of Draco.

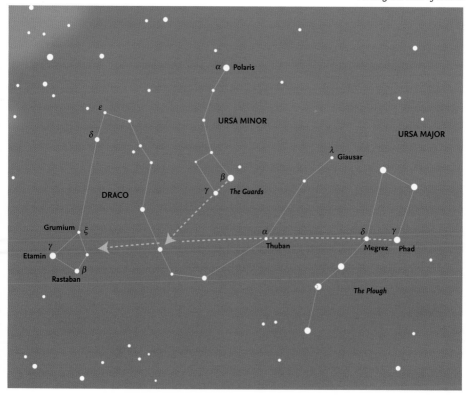

Introduction to the month-by-month guide

The monthly charts

The pages devoted to each month contain a pair of charts showing the appearance of the night sky, looking north and looking south. The charts (as with all the charts in this book) are drawn for the latitude of 50°N, so observers farther north will see slightly more of the sky on the northern horizon, and slightly less on the southern. These areas are, of course, those most likely to be affected by poor observing conditions caused by haze, mist or smoke. In addition, stars close to the horizon are always dimmed by atmospheric absorption, so sometimes the faintest stars marked on the charts may not be visible.

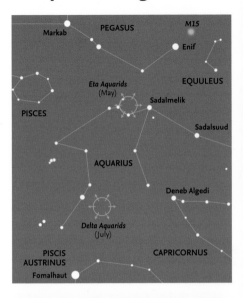

The three times shown for each chart require a little explanation. The charts are drawn to show the appearance at 23:00 GMT for the 1st of each month. The same appearance will apply an hour earlier (22:00 GMT) on the 15th, and yet another hour earlier (21:00 GMT) at the end of the month (shown as the 1st of the following month). GMT is identical to the Universal Time (UT), used by astronomers around the world. In Europe, Summer Time is introduced in March, so the March charts apply to 23:00 GMT on March 1, 22:00 GMT on March 15, but 22:00 BST (British Summer Time) on April 1. The change back from Summer Time (in Europe) occurs in October, so the charts for that month apply to 00:00 BST for October 1, 23:00 BST for October 15, and 21:00 GMT for November 1.

The charts may be used for earlier or later times during the night. To observe two hours earlier, use the charts for the preceding month; for two hours later, the charts for the next month.

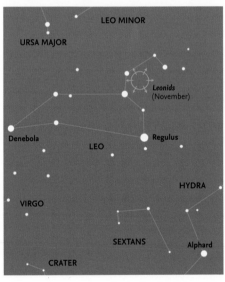

Meteors

The details of specific meteor showers are given in the months in which they come to maximum, regardless of whether they begin or end in other months. It should be noted that not all the respective radiants will be marked on the charts for that particular month, because the radiants may be below the horizon, or lie in constellations that are not readily visible during the month of maximum. For this reason, special charts for the Eta and Delta Aquarids (May and July, respectively)

Shower	Dates of activity 2015	Date of maximum 2015	Possible hourly rate
Quadrantids	January 1–6	January 3–4	70
April Lyrids	April 16–25	April 21–22	< 15
Eta Aquarids	April 19 to May 26	May 5–6	35
Alpha Capricornids	July 11 to August 10	July 28–29	5
Perseids	July 13 to August 26	August 11–12	80
Delta Aquarids	July 21 to August 23	July 27–28	< 20
Alpha Aurigids	August to October	August 28 & September 15	10
Southern Taurids	September 7 to November 19	October 8–9	< 5
Northern Taurids	October 19 to December 10	November 12–13	< 5
Orionids	October 4 to November 14	October 21–22	15–30
Leonids	November 5–30	November 17–18	< 15
Geminids	December 4–16	December 13–14	100+
Ursids	December 17–23	December 21–22	< 10

and the Leonids (November) are given here. As explained earlier, however, meteors from such showers may still be seen, because the most effective region for seeing meteors is some 40–45° away from the radiant, and that area of sky may well be above the horizon. For convenience, a table of the best meteor showers visible during the year is also given here.

The photographs

As an aid to identification – especially as some people find it difficult to relate charts to the actual stars they see in the sky – one or more photographs of constellations visible in certain specific months are included. It should be noted, however, that because of the limitations of the photographic and printing processes, and the differences between the sensitivity of different individuals to faint starlight (especially in their ability to detect different colours), and the degree to which they have become adapted to the dark, the apparent brightness of stars in the photographs will not necessarily precisely match that seen by any one observer.

The Moon calendar

The Moon calendar is largely self-explanatory. It shows the phase of the Moon for every day of the month, with the exact times (in Universal Time) of New Moon, First Quarter, Full Moon, and Last Quarter. Because the times are calculated from the Moon's actual orbital parameters, some of the times shown will, naturally, fall during daylight, but any difference is too small to affect the appearance of the Moon on that date.

The Moon

The section on the Moon includes details of any lunar or solar eclipses that may occur during the month (visible from anywhere on Earth). Similar information is given about any important occultations. Mainly, however, this section summarizes when the Moon passes close to planets or the five prominent stars close to the ecliptic. The dates when the Moon is closest to the Earth (at perigee) and farthest from it (at apogee) are shown in the monthly calendars, and only mentioned here when they are particularly significant, such as the nearest and farthest during the year.

The Planets

Brief details are given of the location, movement and brightness of the planets from Mercury to Saturn throughout the month. None of the planets can, of course, be seen when they are close to the Sun, so such periods are generally noted. All of the planets may sometimes lie on the opposite side of the Sun to the Earth (at what is known as superior

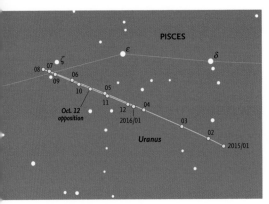

The path of Uranus in 2015. Uranus comes to opposition on October 12.

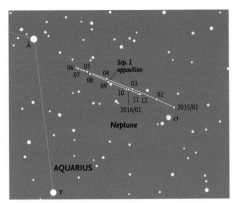

The path of Neptune in 2015. Neptune comes to opposition on September 1.

conjunction), but in the case of the inferior planets, Mercury and Venus, they may also pass between the Earth and the Sun (at inferior conjunction) and are normally invisible for a longer or shorter period of time. Those two planets are normally easiest to see around either eastern or western elongation, in the evening or morning sky, respectively. Not every such elongation is favourable, however, so although every elongation is listed, only those where observing conditions are most favourable are shown in the individual diagrams of events.

The dates at which the superior planets reverse their motion (from direct motion to retrograde, and retrograde to direct) and of opposition (when a planet generally reaches its maximum brightness) are given. Some planets, especially distant Saturn, may spend most or all of the year in a single constellation. Jupiter and Saturn are normally easiest to see around opposition which occurs every year. Mars, by contrast, moves relatively rapidly against the background stars and in some years never comes to opposition.

Uranus is not included in the monthly details because it is generally at the limit of naked-eye visibility (magnitude 5.7–5.9), although bright enough to be visible in binoculars, or even with the naked eye under exceptionally dark skies. Its path in 2015 is shown on the special chart above. It comes to opposition on October 12 in the constellation of Pisces. New Moon occurs one day later, so the planet, at magnitude 5.7, should be readily detectable for some days on either side of that date, when its change in position relative to the background stars will become noticeable.

Somewhat similar considerations apply to Neptune, although this is always fainter (magnitude 7.9–8.0 in 2015), but still visible in most binoculars. It reaches opposition on September 1 in Aquarius, but the Moon, approaching Last Quarter in the neighbouring constellation of Pisces, may cause some interference. Again, Neptune's path in 2015 and its position at opposition are shown in the chart above.

The ecliptic charts

Although the ecliptic charts are primarily designed to show the positions and motions of the major planets, they also show the motion of the Sun during the month. The light-tinted area shows the area of the sky that is invisible during daylight, but the darker area gives an indication of which constellations are likely to be visible at some time of the night.

The closer a planet is to the border between dark and light, the more difficult it will be to see in the twilight. (If it is in the light area, of course, it will be invisible, drowned by the light from the Sun.)

The monthly calendar

For each month, a calendar shows details of significant events, including when planets are close to one another in the sky, close to the Moon, or close to any one of five bright stars that are spaced along the ecliptic. The times shown are given in Universal Time (UT), always used by astronomers throughout the year, and which is identical to Greenwich Mean Time (GMT). So, even during the summer months, they do not show Summer Time, which will always be one hour later than the time shown.

The diagrams of interesting events

Each month, a number of diagrams show the appearance of the sky when certain events take place. However, the exact positions of celestial objects and their separations greatly depend on the observer's position on Earth. When the Moon is one of the objects involved, because it is relatively close to Earth, there may be very significant changes from one location to another. Close approaches between planets or between a planet and a star are less affected by changes of location, which may thus be ignored.

The diagrams showing the appearance of the sky are drawn for the latitude of London, so will be approximately correct for most of Britain and Europe. However, for an observer farther north (say Edinburgh), a planet or star listed as being north of the Moon will appear even farther north, whereas one south of the Moon will appear closer to it – or may even be hidden (occulted) by it. For an observer farther south than London, there will be corresponding changes in the opposite direction: for a star or planet south of the Moon the separation will increase, and for one north of the Moon the separation will decrease. For example, on 5 September 2015 during the occultation of Aldebaran (page 73), the apparent path of the star appears farther north seen from Edinburgh than from London.

Ideally, details should be calculated for each individual observer, but this is obviously impractical. In fact, positions and separations are actually calculated for a theoretical observer located at the centre of the Earth.

So the details given regarding the positions of the various bodies should be used as a guide to their location. A similar situation arises with the times that are shown. These are calculated according to certain technical criteria, which need not concern us here. However, they do not necessarily indicate the exact time when two bodies are closest together. Similarly, dates and times are given, even if they fall in daylight, when the objects are likely to be completely invisible. However, such times do give an indication that the objects concerned will be in the same general area of the sky during both the preceding, and the following nights.

Key to the symbols used on the monthy star maps.

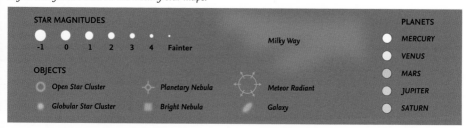

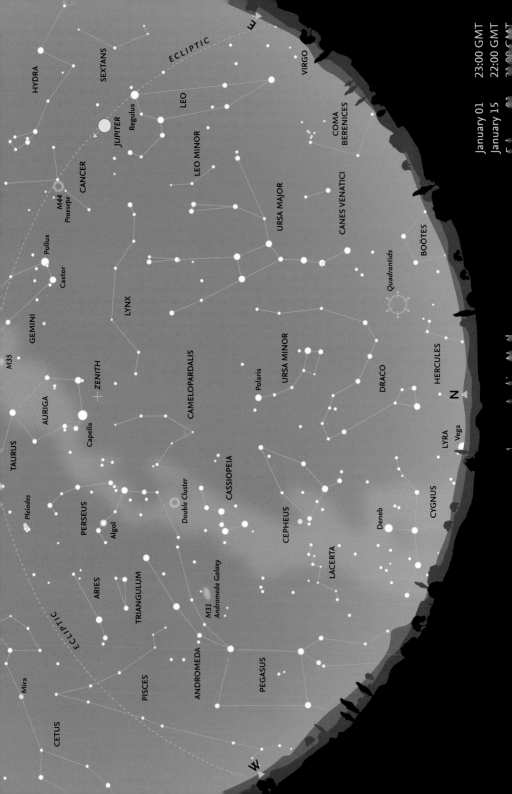

HYDRA

SEXTANS

ECLIPTIC

VIRGO

LEO

Regulus

JUPITER

COMA
BERENICES

CANCER

M44
Praesepe

LEO MINOR

Pollux

URSA MAJOR

Castor

CANES VENATICI

GEMINI

LYNX

BOÖTES

Quadrantids

M35

ZENITH

URSA MINOR

DRACO

HERCULES

CAMELOPARDALIS

Polaris

N

AURIGA

Capella

LYRA

Vega

TAURUS

CASSIOPEIA

Double Cluster

CYGNUS

Pleiades

CEPHEUS

PERSEUS

Algol

Deneb

ARIES

LACERTA

TRIANGULUM

M31
Andromeda Galaxy

ANDROMEDA

PEGASUS

PISCES

ECLIPTIC

Mira

CETUS

W

January – Looking North

Most of the important circumpolar constellations are easy to see in the northern sky at this time of year. **Ursa Major** stands more-or-less vertically above the horizon in the northeast, with the zodiacal constellation of **Leo** rising in the east. To the north, the stars of **Ursa Minor** lie below Polaris (the Pole Star). The head of **Draco** is low on the northern horizon, but may be difficult to see unless observing conditions are good. Both **Cepheus** and **Cassiopeia** are readily visible in the northwest, and even the faint constellation of **Camelopardalis** is high enough in the sky for it to be easily visible. **Auriga**, with bright **Capella**, is high overhead near the zenith, while **Perseus** and **Andromeda** are visible farther down towards the west, where the Great Square of **Pegasus** is approaching the horizon.

Meteors

One of the strongest and most consistent meteor showers occurs in January: the Quadrantids, which are visible January 1–6, with maximum on January 3–4. They are brilliant, bluish and yellowish-white meteors and at maximum may reach a rate of 70 meteors per hour. Unfortunately in 2015, Full Moon occurs on January 5, which will greatly interfere with visibility. The parent object is minor planet 2003 EH.

They are named after the former constellation of **Quadrans Muralis** (the Mural Quadrant), an early form of astronomical instrument. The Quadrantid meteor radiant is now within the northernmost part of **Boötes**, roughly half-way between θ Boötis and τ Herculis.

Minor planet

On January 29, minor planet **(3) Juno** reaches opposition at mag. 8.1 passing near the distinctive asterism (or group of stars), known as the 'Head of Hydra'. The chart covers the period (including the end of 2014) when Juno is brighter than about magnitude 9.0 and visible under good conditions with binoculars. The area included on this detailed chart (showing all stars brighter than mag. 8.6) is indicated on the 'looking south' chart overleaf. (On April 26, Juno will be occulted by the Moon, but this will be visible only from the Pacific.)

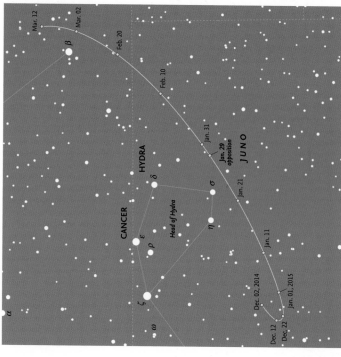

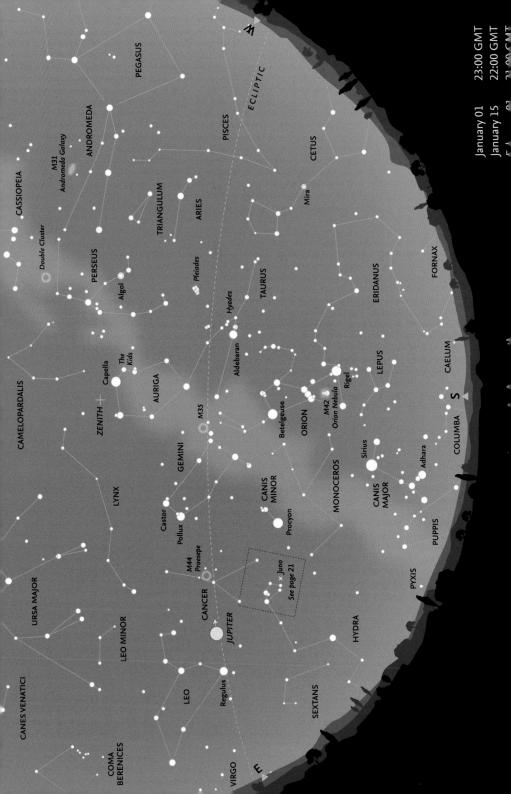

CANES
BERENICES

COMA
BERENICES

CANES VENATICI

URSA MAJOR

LEO MINOR

LYNX

LEO

Regulus

CAMELOPARDALIS

Castor
Pollux

GEMINI

M44
Praesepe

CANCER

JUPITER

ZENITH

Capella

The
Kids

AURIGA

M35

CASSIOPEIA

Double Cluster

PERSEUS

Algol

Pleiades

Hyades

TRIANGULUM

ARIES

TAURUS

Aldebaran

Betelgeuse

ORION

M42
Orion Nebula

Rigel

CANIS
MINOR

Procyon

Juno

See page 21

MONOCEROS

Sirius

CANIS
MAJOR

Adhara

LEPUS

ERIDANUS

FORNAX

CAELUM

COLUMBA

S

PUPPIS

PYXIS

HYDRA

SEXTANS

E

VIRGO

PEGASUS

ANDROMEDA

M31
Andromeda Galaxy

PISCES

ECLIPTIC

CETUS

Mira

N

January – Looking South

At this time of year the southern sky is dominated by **Orion**. This is the most prominent constellation during the winter months, when it is visible at some time during the night. (A photograph of Orion appears on page 87.) It has a highly distinctive shape, with a line of three stars that form the 'Belt'. To most observers, the bright star at the northwestern corner of the constellation, **Betelgeuse** (α Orionis), shows a reddish tinge, in contrast to the brilliant bluish-white colour of the bright star at the southwestern corner, **Rigel** (β Orionis). The three stars of the belt lie across the celestial equator. A vertical line of three 'stars' forms the 'Sword' that hangs to the south of the Belt. With good viewing conditions, the central 'star' appears as a hazy spot, even to the naked eye. This is actually the Orion Nebula. Binoculars will reveal the four stars of the Trapezium, which illuminate the nebula.

The line of Orion's Belt points up to the northwest towards **Taurus** (the Bull) and orange-tinted **Aldebaran** (α Tauri). Close to Aldebaran, there is a conspicuous 'V' of stars, pointing down to the southwest, called the **Hyades** cluster. (Despite appearances, Aldebaran

is not part of the cluster.) Farther along, the same line from Orion takes you close to a bright cluster of stars, the **Pleiades**, or Seven Sisters. Even the smallest pair of binoculars reveals this cluster as a beautiful group of bluish-white stars. The two most conspicuous of the other stars in Taurus lie directly above Orion, and form an elongated triangle with Aldebaran. The northernmost, β Tauri, was once considered to be part of the constellation of Auriga.

Also above Orion at this time of year is the constellation of **Auriga** (the Charioteer), with brilliant **Capella** (α Aurigae), close to the zenith, directly overhead. Slightly to the west of Capella lies a small triangle of fainter stars, known as 'The Kids'. (Ancient mythological representations of Auriga show him carrying two young goats.) Together with that northernmost bright star in Taurus, β Tauri, the body of Auriga forms a large pentagon on the sky, with the Kids lying on the western side.

The Moon's phases for January

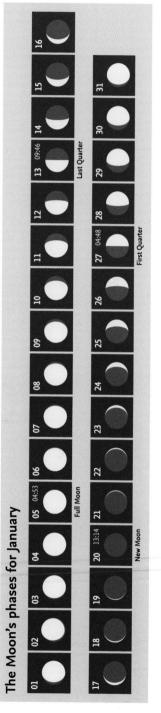

January – Moon and Planets

The Earth

The Earth reaches perihelion (the closest point to the Sun in its annual orbit) on 4 January 2015, at 06:36 Universal Time. Its distance is then 0.9833 AU (147,096,134 km).

The Moon

On January 2, the Moon appears southeast of **Aldebaran**. (Closest approach is in daylight, earlier in the day.) On January 8 it passes near **Jupiter** (in **Leo**) and, a day later, it is south of Regulus (α Leonis). On January 13, it may be seen north of **Spica** (α Virginis). On January 23, the waxing crescent passes north of **Mars** in the western sky.

The Planets

Jupiter, which is retrograding (moving westwards) in **Leo** (near the border with **Cancer**) is prominent throughout the night, declining from magnitude -2.4 to -2.6 over the month. On January 8 it is close to the waning gibbous Moon, but closest approach occurs at 08:22 in daylight. **Mercury** comes to greatest eastern elongation on January 14, low on the western horizon and not easy to see although **Venus** is nearby at magnitude -3.9. Venus gradually becomes higher in the sky and more visible as the month progresses, remaining at the same magnitude. Both it and **Mars** (magnitude 1.1–1.2) are in **Aquarius**, visible in the western sky after sunset. **Saturn** is slightly brighter than Mars at magnitude 0.6–0.5. However it is in **Libra**, near the border with **Scorpius**, and therefore rises well after midnight (around 04:00).

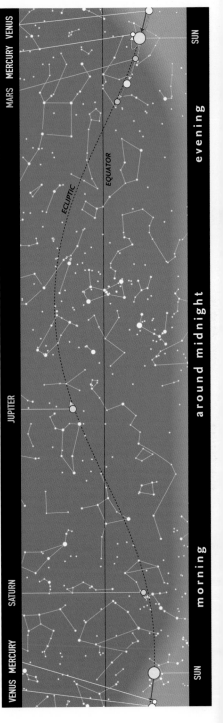

The path of the Sun and the planets along the ecliptic in January.

Calendar for January

01–06		Quadrantid meteor shower
02	11:59	Aldebaran 1.4°S of Moon
03–04		Quadrantid shower maximum
04	06:36	Earth at perihelion (147,096,134 km = 0.9833 AU)
05	04:53	Full Moon
06	02:17	Pollux 11.8°N of Moon
08	08:22	Jupiter 5.1°N of Moon
09	02:00	Regulus 4.1°N of Moon
09	18:18	Moon at apogee
13	09:46	Last Quarter
13	09:51	Spica 3.0°S of Moon
14	20:31	Mercury greatest elongation (18.9° E, mag. -0.7)
16	11:32	Saturn 1.9°S of Moon
16	23:24	Antares 8.8°S of Moon
20	13:14	New Moon
21	17:39	Mercury 3.0°S of Moon
21	20:07	Moon at perigee
22	04:59	Venus 5.6°S of Moon
23	04:40	Mars 3.9°S of Moon
25	11:51	Uranus 0.6°S of Moon
27	04:48	First Quarter
29	17:31	Aldebaran 1.2°S of Moon
29	22:44	Juno at opposition (mag. 8.1)
30	13:45	Mercury inferior conjunction

Morning 7:00

January 8–9 • The Moon with Jupiter and Regulus in the morning sky.

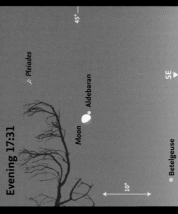

Morning 7:00

January 16 • The Moon and Saturn and Antares in the morning sky.

Evening 17:30

January 21–23 • The Moon passes Mercury, Venus and Mars, after sunset. Mercury is very close to the horizon.

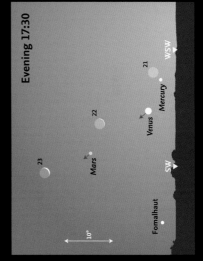

Evening 17:31

January 29 • The Moon close to Aldebaran, high in the southeastern sky.

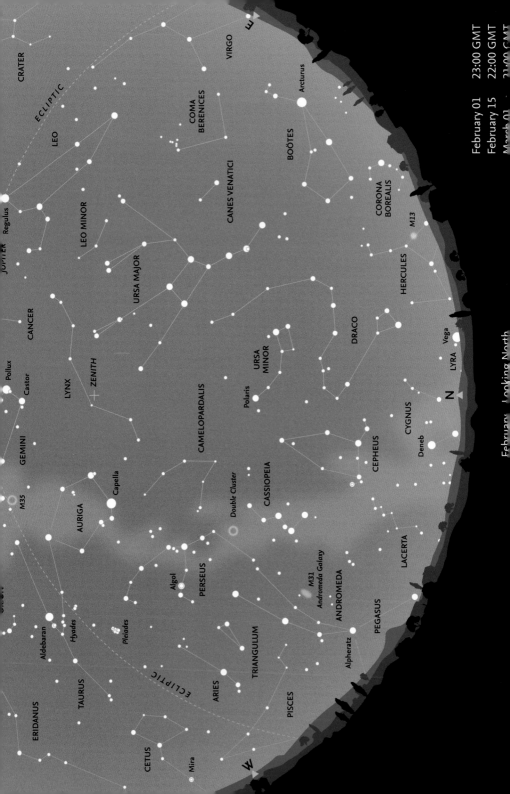

CRATER

ECLIPTIC

LEO

VIRGO

COMA BERENICES

Arcturus

BOÖTES

Regulus

JUPITER

LEO MINOR

CANES VENATICI

CORONA BOREALIS

M13

URSA MAJOR

CANCER

HERCULES

ZENITH

LYNX

CAMELOPARDALIS

URSA MINOR

DRACO

Pollux

Castor

Polaris

Vega

GEMINI

LYRA

CYGNUS

N

M35

CEPHEUS

Deneb

Capella

Double Cluster

AURIGA

CASSIOPEIA

LACERTA

Algol

PERSEUS

M31
Andromeda Galaxy

ANDROMEDA

PEGASUS

Hyades

Aldebaran

Pleiades

TRIANGULUM

Alpheratz

TAURUS

ERIDANUS

ARIES

PISCES

CETUS

Mira

ECLIPTIC

W

February, Looking North

February – Looking North

The months of January and February are probably the best time for seeing the section of the Milky Way that runs in the northern and western sky from **Cygnus**, low on the northern horizon, through **Cassiopeia**, **Perseus** and **Auriga** and then down through **Gemini** and **Orion**. Although not as readily visible as the denser star clouds of the summer Milky Way, on a clear night so many stars may be seen that even a distinctive constellation such as **Cassiopeia** is not immediately obvious.

The head of **Draco** is now higher in the sky and easier to recognize. **Deneb** (α Cygni), the brightest star in **Cygnus**, may just be visible almost due north at midnight, early in the month, if the sky is very clear and the horizon clear of obstacles. **Vega** (α Lyrae) in **Lyra** is so low that it is difficult to see, but may become visible later in the night. The constellation of **Boötes** – sometimes described as shaped like a kite, an ice-cream cone or the letter 'P' – with orange-tinted **Arcturus** (α Boötis), is beginning to clear the eastern horizon. Arcturus, at magnitude –0.05, is the brightest star north of the celestial equator. The inconspicuous constellation of **Coma Berenices** is now well above the horizon in the east. The concentration of faint stars at the northeastern corner somewhat resembles a tiny, detached portion of the Milky Way.

On the other side of the sky, in the northwest, most of the constellation of **Andromeda** is still easily seen, although **Alpheratz** (α Andromedae), the star that forms the northeastern corner of the Great Square of **Pegasus** – even though it is actually part of Andromeda – is becoming close to the horizon and more difficult to detect. High overhead, at the zenith, try to make out the very faint constellation of **Lynx**. It was introduced in 1687 by the famous astronomer Johannes Hevelius to fill the largely blank area between **Auriga**,

Gemini and **Ursa Major**, and is reputed to be so named because one needed the eyes of a lynx to detect it.

The Dawn spaceprobe

In February, NASA's Dawn spaceprobe is due to arrive at, and go into orbit around, the dwarf planet **(1) Ceres**. The earlier of Dawn's two main objectives was to orbit and survey the minor planet **(4) Vesta** between July 2011 and July 2012. There it was extremely successful, providing striking images of the surface and detailed information about the body's complex geology.

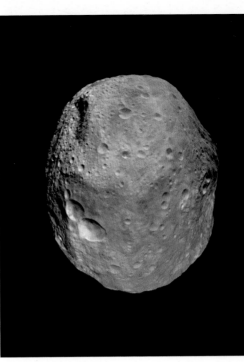

A mosaic of images of Vesta obtained by the Dawn spaceprobe.

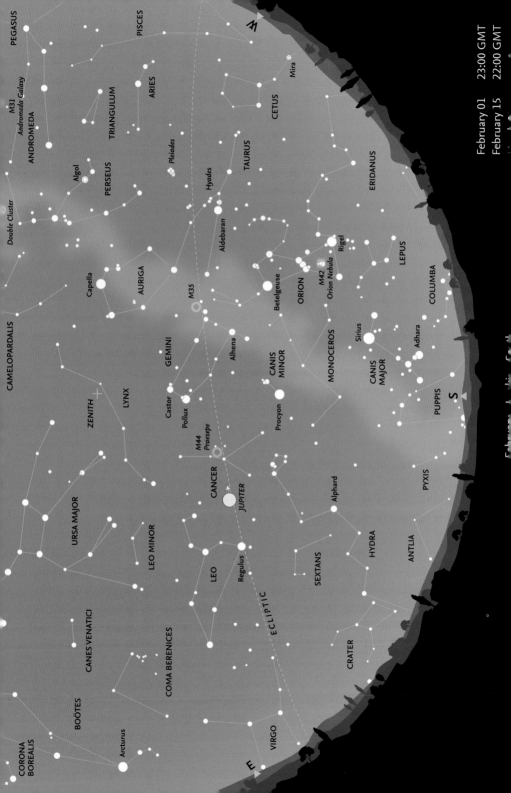

February Looking South

February 01   23:00 GMT
February 15   22:00 GMT

February – Looking South

Apart from Orion, the most prominent constellation visible this month is **Gemini**, with its two lines of stars running southeast towards Orion. Many people have difficulty in remembering which is which of the two stars **Castor** and **Pollux**. Think of them in alphabetical order: Castor (α Geminorum), the fainter star, is closer to the North Celestial Pole. Pollux (β Geminorum) is the brighter of the two, but is farther away from the Pole. Castor is remarkable because it is actually a multiple system, consisting of no less than six individual stars.

Using Orion's belt as a guide, it points down to the southeast towards **Sirius**, the brightest star in the sky (at magnitude −1.4) in the constellation of **Canis Major**, the whole of which is now clear of the southern horizon. Forming a triangle with **Betelgeuse** in Orion and **Sirius** in Canis Major is **Procyon**, the brightest star in the small constellation of **Canis Minor**. Between Canis Major and Canis Minor is the faint constellation of **Monoceros**, which actually straddles the Milky Way, which is difficult to see in this area. Directly east of Procyon is the highly distinctive asterism of six stars that form the 'head' of **Hydra**, the largest of all 88 constellations, and which trails such a long way across

the sky that it is only in mid-March around midnight that the whole constellation becomes visible.

The constellation of Gemini. The two brightest stars are Castor and Pollux and can be found in the left part of the photograph.

The Moon's phases for February

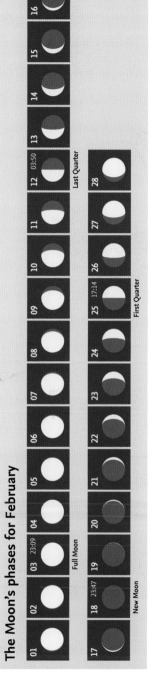

February – Moon and Planets

The Moon

On February 4, one day past Full, the Moon appears near *Jupiter*. (Closest approach occurs during daylight.) Jupiter is then right on the boundary between *Leo* and *Cancer*. At the same time (08:42) the next day it is closest to Regulus (α Leonis), also, of course, in daylight. On February 12–13 at a waning gibbous phase, just after Last Quarter, the Moon passes close to *Saturn* (Feb. 12) and then *Antares* (Feb. 13). On February 25, a few hours after First Quarter, there is an appulse (close approach) to *Aldebaran* in *Taurus*.

The Planets

Mercury comes to western elongation on February 24, but is too low on the dawn horizon to be readily visible. At the beginning of the month, *Venus* is in the western sky (in *Aquarius*) at magnitude -3.9, setting about 90 minutes after sunset. It gradually climbs higher in the sky throughout the month, remaining at magnitude -3.9. *Mars* (magnitude 1.2–1.3) is initially also in Aquarius, higher in the sky than Venus, but by the end of the month is slightly lower, both planets being then in *Pisces*. *Jupiter* is still retrograding on the very border of *Leo* and Cancer. (It is just within Cancer at opposition on February 6, at magnitude -2.6.)

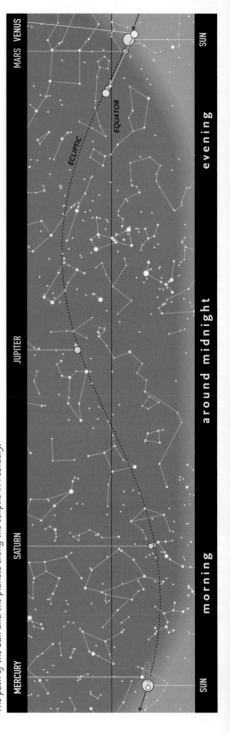

The path of the Sun and the planets along the ecliptic in February.

Calendar for February

03	23:09	Full Moon
04	08:42	Jupiter 5.2°N of Moon
05	08:42	Regulus 4.0°N of Moon
06	06:26	Moon at apogee
06	18:20	Jupiter at opposition (mag. -2.6)
09	16:48	Spica 3.3°S of Moon
12	03:50	Last Quarter
12	23:49	Saturn 2.1°S of Moon
13	08:35	Antares 9.0°S of Moon
17	06:20	Mercury 3.5°S of Moon
18	23:47	New Moon
19	07:28	Moon at perigee
20–21		Moon/Mars/Venus
21	00:55	Venus 2.0°S of Moon
21	01:29	Mars 1.5°S of Moon
21	22:16	Uranus 0.3°S of Moon
24	16:23	Mercury greatest elongation (26.7° W, mag. 0.0)
25	17:14	First Quarter
25	23:26	Aldebaran 1.0°S of Moon.

Morning 3:00

February 1–5 • *The Moon passes Alhena (in Gemini), Jupiter and Regulus. Castor and Pollux, Procyon and Alphard (in Hydra) are also near.*

Morning 6:30

February 13 • *The Moon with Saturn and Antares in the morning sky.*

Evening 18:30

February 20–21 • *The Moon with Mars and Venus in the western sky, after sunset.*

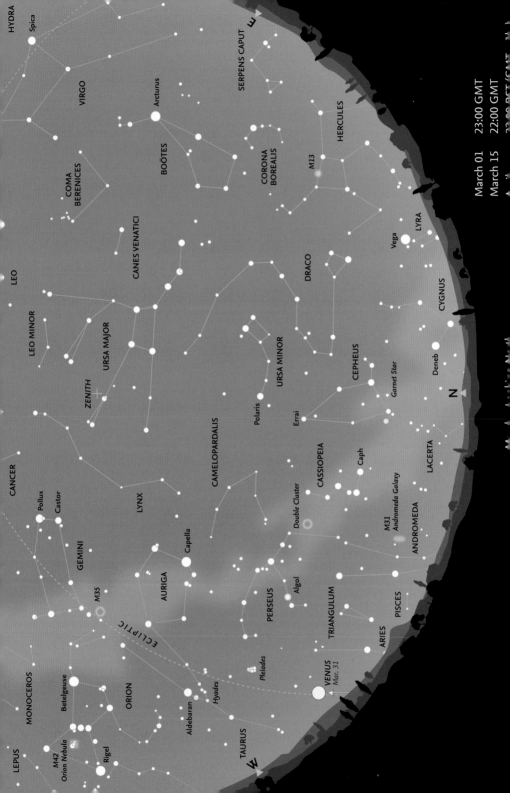

HYDRA
Spica

VIRGO

SERPENS CAPUT

Arcturus

BOÖTES

COMA
BERENICES

CORONA
BOREALIS

HERCULES

M13

CANES VENATICI

LEO

LYRA

Vega

LEO MINOR

DRACO

URSA MAJOR

CYGNUS

ZENITH

URSA MINOR

CEPHEUS

Deneb

N

CANCER

CAMELOPARDALIS

Polaris

Garnet Star

Errai

Pollux

Castor

LYNX

CASSIOPEIA

Caph

LACERTA

GEMINI

Capella

Double Cluster

M31
Andromeda Galaxy

AURIGA

ANDROMEDA

M35

PERSEUS

Algol

TRIANGULUM

PISCES

MONOCEROS

Betelgeuse

ECLIPTIC

ARIES

Pleiades

VENUS
Mar. 31

ORION

Aldebaran

Hyades

LEPUS

M42
Orion Nebula

Rigel

TAURUS

W

March – Looking North

In March, the Sun crosses the celestial equator on Friday, March 20, at the vernal equinox, when day and night are of almost exactly equal length, and the season of spring is considered to have begun. (The hours of daylight and darkness change most rapidly around the equinoxes, in March and September.) It is also in March that Summer Time begins in Europe (on Sunday, March 29) so the charts show the appearance at 23:00 GMT for March 1 and 22:00 BST for April 1. Early in the month, the constellation of *Cepheus* lies almost due north, with the distinctive 'W' of *Cassiopeia* to its west. Cepheus lies across the border of the Milky Way and is often described as like the gable end of a house or a church tower and steeple. Despite the large number of stars revealed at the base of the constellation by binoculars, one star stands out because of its deep red colour. This is Mu (μ) Cephei, also known as the *Garnet Star* because of its striking colour. It is a truly gigantic star, a red supergiant, and one of the largest stars known. It is at least 1000 times the diameter of the Sun, and if placed in the Solar System would extend to about the orbit of Jupiter. (Betelgeuse, in Orion, is also a red supergiant, and is perhaps 950 times the diameter of the Sun.)

Below Cepheus to the east (to the right), it may be possible to catch a glimpse of *Deneb* (α Cygni), just above the horizon. Slightly farther round towards the northeast, *Vega* (α Lyrae) is marginally higher in the sky. From southern Britain, Deneb is just far enough north to be circumpolar (although difficult to see in January and February because it is so low on the northern horizon). Vega, by contrast, farther south, is completely hidden during the depths of winter.

Total solar eclipse of March 20

On March 20 there is a total solar eclipse with the path of totality passing across the northern Atlantic Ocean. As shown in the diagram,

only from the Faeroe islands, lying between Iceland and northern Scotland, and from most of the archipelago of Svalbard, will the eclipse be total. Greatest eclipse (southeast of Iceland) is at 09:46 UT. A large partial eclipse will be visible from all of the British Isles. At London, the eclipse begins at 08:25, maximum eclipse (85 per cent obscured) at 09:31, and the eclipse ends at 10:41. At Edinburgh the respective times are: start 08:30, maximum 09:35 (93 per cent obscured), and end 10:43. This eclipse is unusual. Because it occurs on the day of the spring equinox, it is actually visible at the North Pole.

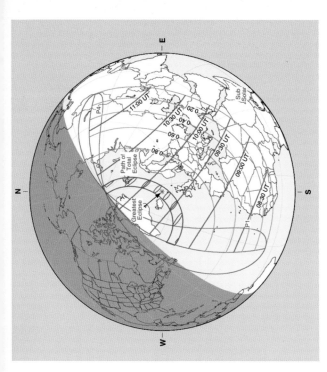

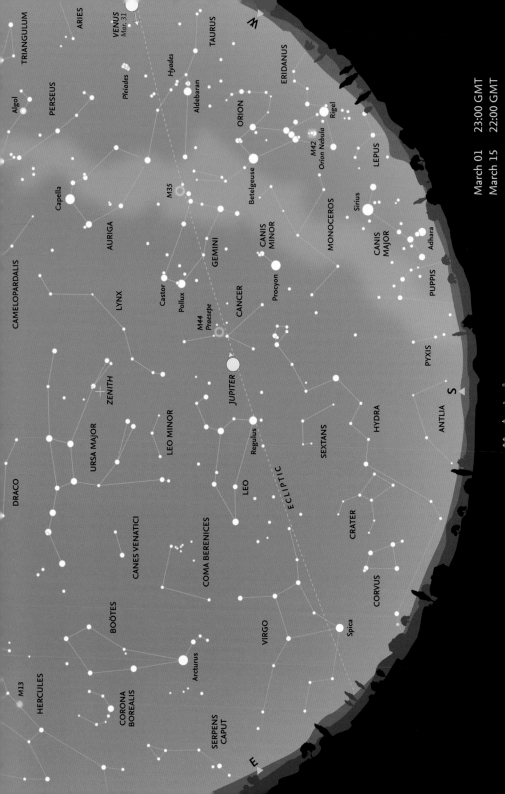

TRIANGULUM

ARIES

VENUS
Mar. 31

TAURUS

Algol

PERSEUS

Hyades

Pleiades

Aldebaran

ERIDANUS

ORION

Rigel

M42
Orion Nebula

Betelgeuse

LEPUS

M35

Capella

AURIGA

GEMINI

CANIS
MINOR

Sirius

MONOCEROS

CANIS
MAJOR

Adhara

CAMELOPARDALIS

Castor

Pollux

CANCER

Procyon

PUPPIS

LYNX

M44
Praesepe

PYXIS

ZENITH

URSA MAJOR

LEO MINOR

JUPITER

HYDRA

S

ANTLIA

DRACO

Regulus

SEXTANS

LEO

ECLIPTIC

CANES VENATICI

CRATER

COMA BERENICES

CORVUS

BOÖTES

VIRGO

Arcturus

Spica

M13

HERCULES

CORONA
BOREALIS

SERPENS
CAPUT

E

March 01   23:00 GMT
March 15   22:00 GMT

March – Looking South

Due south at 22:00 at the beginning of the month, lying between the constellations of **Gemini** in the west and **Leo** in the east, and fairly high in the sky above the head of **Hydra**, is the faint and rather undistinguished zodiacal constellation of **Cancer**. Rather like the triskelion, the symbol for the Isle of Man, it has three 'legs' radiating from the centre, where there is an open cluster, M44 or **Praesepe** ('the Manger' but also known as 'the Beehive'). On a clear night this is just a hazy spot to the naked eye, but appears in binoculars as a group of dozens of individual stars.

Also prominent in March is the constellation of **Leo**, with the 'backward question mark' (or 'Sickle') of bright stars forming the head of the mythological lion. **Regulus** (α Leonis) – the 'dot' of the 'question mark' or the handle of the sickle and the brightest star in Leo – lies very close to the ecliptic and is one of the few first-magnitude stars that may be occulted by the Moon. The next occultation occurs on 18 December 2016, visible only from southern Australia and part of Antarctica. A series of occultations follows in 2017, but not until 25 July 2017 and 8 December 2017 will occultations be visible from any part of Europe.

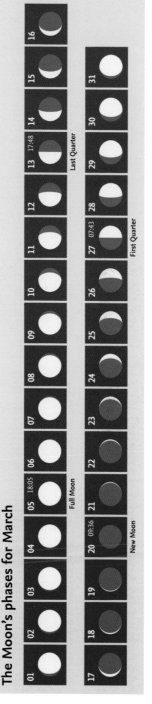

The faint zodiacal constellation of Cancer lies between Castor and Pollux (on the right) and Regulus (on the left). To its south is the distinctive asterism of six stars marking the head of Hydra.

The Moon's phases for March

01	02	03	04	05 18:05	06
				Full Moon	

07	08	09	10	11	12	13 17:48	14	15	16
						Last Quarter			

17	18	19	20 09:36	21	22	23	24	25	26	27 07:43	28
			New Moon							**First Quarter**	

29	30	31

March – Moon and Planets

The Moon

On March 3, the waxing gibbous Moon passes south of *Jupiter* and, the next day, south of *Regulus* (α Leonis). A similar pair of appulses occurs on March 30 and 31. On March 8 it is north of *Spica* (α Virginis) and some hours later (in daylight) passes north of *Antares* (α Scorpii). It passes just north of *Saturn* on March 12 and Aldebaran on March 25. A total solar eclipse (partial from the British Isles) occurs on March 20 (see diagram page 33).

The Planets

Mercury is too low and close to the Sun to be visible in March. *Venus* begins the month in *Pisces*, rising higher and moving into *Aries* as it brightens slightly from magnitude -3.9 to -4.0. It sets about 22:00 at the end of the month. *Mars* may be seen in the west after sunset below Venus, but it becomes lost in evening twilight by the end of the month. *Jupiter* is still retrograding in *Cancer*, brightening slightly from magnitude -2.5 to -2.3 over the month. It initially sets at about 06:00, and towards 04:00 at the end of March. *Saturn* (in *Scorpius*) rises in the southeast shortly after midnight and brightens slightly (from magnitude 0.5 to 0.3) over the month.

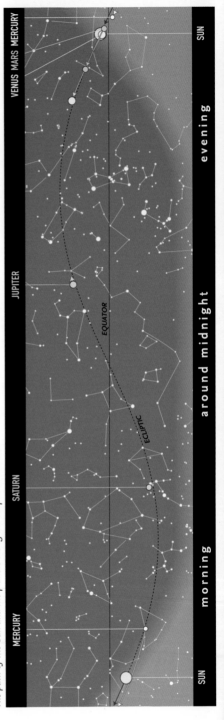

The path of the Sun and the planets along the ecliptic in March.

Calendar for March

03	08:03	Jupiter 5.5°N of Moon
04	04:54	Regulus 4.0°N of Moon
05	07:33	Moon at apogee
05	18:05	Full Moon
08		Daylight Saving Time begins (North America)
08	22:46	Spica 3.5°S of Moon
12	08:03	Saturn 2.2°S of Moon
12	05:39	Antares 9.2°S of Moon
13	17:48	Last Quarter
19	04:56	Mercury 5.2°S of Moon
19	19:38	Moon at perigee
20	09:36	New Moon
20	09:46	Solar eclipse
20	22:45	Spring equinox
21	11:18	Uranus 0.1°S of Moon
21	22:14	Mars 1.0°N of Moon
22	19:50	Venus 2.8°N of Moon
25	07:17	Aldebaran 0.88S° of Moon
27	07:43	First Quarter
29		Summer Time begins (Europe)
30	10:26	Jupiter 5.6°N of Moon
31	21:11	Regulus 4.0°N of Moon

Evening 22:00

March 2–4 • The Moon passes Jupiter and Regulus.

Evening 19:00

March 21–22 • The Moon with Mars and Venus in the evening sky.

Morning 5:30

March 8–13 • The Moon passes Spica, Saturn and Antares.

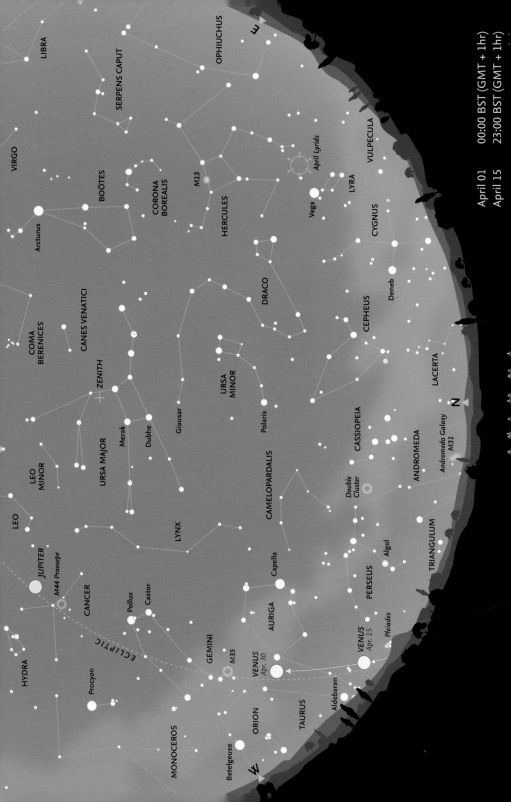

LIBRA

OPHIUCHUS

E

SERPENS CAPUT

VIRGO

April Lyrids

VULPECULA

BOÖTES

M13

LYRA

CORONA
BOREALIS

HERCULES

Vega

CYGNUS

Arcturus

Deneb

CANES VENATICI

DRACO

CEPHEUS

COMA
BERENICES

ZENITH

LACERTA

URSA
MINOR

Merak

N

Dubhe

Giausar

Polaris

CASSIOPEIA

ANDROMEDA

URSA MAJOR

Andromeda Galaxy

LEO
MINOR

CAMELOPARDALIS

M31

LEO

Double
Cluster

LYNX

ANDROMEDA

JUPITER

Algol

TRIANGULUM

M44 Praesepe

Capella

CANCER

PERSEUS

AURIGA

Pollux

Castor

ECLIPTIC

GEMINI

VENUS
Apr. 15

Pleiades

HYDRA

M35

VENUS
Apr. 30

Procyon

Aldebaran

MONOCEROS

ORION

TAURUS

Betelgeuse

W

April 01   00:00 BST (GMT + 1hr)
April 15   23:00 BST (GMT + 1hr)

April – Looking North

Cygnus and the brighter regions of the Milky Way are now becoming visible, rising in the northeast, with the small constellation of **Lyra** and the distinctive 'Keystone' of **Hercules** above them. This asterism is very useful for locating the bright globular cluster M13 (see map on page 53). The winding constellation of **Draco** weaves its way from the quadrilateral of stars that marks its 'head', on the border with Hercules, to end at **Giausar** (λ Draconis) between Polaris (α Ursae Minoris) and the 'Pointers', **Dubhe** and **Merak** (α and β Ursae Majoris, respectively). **Ursa Major** is high overhead, near the zenith. **Auriga** is still clearly seen in the northwest, but, by the end of the month, the southern portion of **Perseus** is

starting to dip below the northern horizon. The very faint constellation of **Camelopardalis** lies in the northwest between Polaris and the constellations of **Auriga** and **Perseus**.

Meteors

A minor meteor shower, the **Lyrids**, peaks on April 21 and 22. Although the hourly rate is not very high (about 15 meteors per hour), the meteors are fast and some leave persistent trains. This year, the maximum is three days after New Moon, so moonlight should not cause significant interference. The parent object is comet C/1861 G1 (Thatcher).

The constellation of Lyra is marked by the bright star Vega, which for most of Europe only dips below the northern horizon in mid-winter, with a distinctive quadrilateral of stars to its southeast.

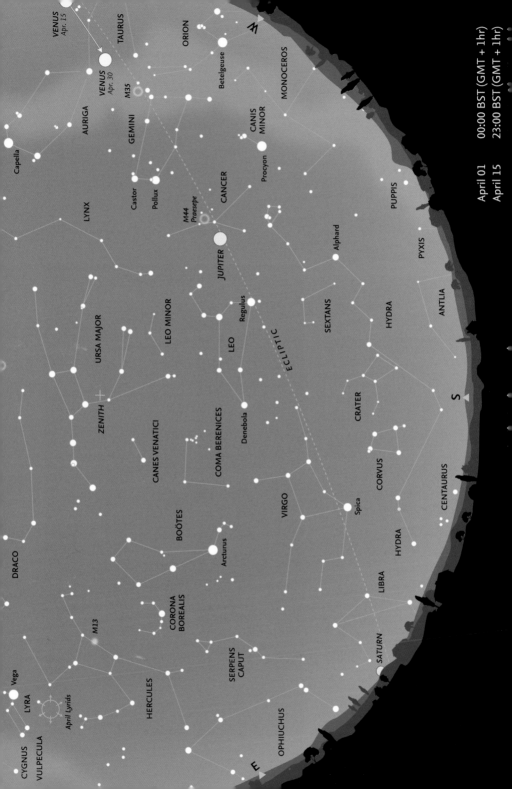

VENUS
Apr. 15

TAURUS

ORION

N

VENUS
Apr. 30

M35

AURIGA

GEMINI

Betelgeuse

MONOCEROS

Capella

CANIS
MINOR

Castor

Pollux

Procyon

CANCER

LYNX

M44
Praesepe

PUPPIS

JUPITER

Alphard

PYXIS

URSA MAJOR

LEO MINOR

SEXTANS

Regulus

ANTLIA

LEO

HYDRA

ECLIPTIC

ZENITH

CANES VENATICI

CRATER

Denebola

DRACO

COMA BERENICES

CORVUS

VIRGO

CENTAURUS

BOÖTES

S

Arcturus

Spica

M13

HYDRA

CORONA
BOREALIS

LIBRA

Vega

SERPENS
CAPUT

LYRA

SATURN

April Lyrids

CYGNUS

HERCULES

VULPECULA

OPHIUCHUS

E

April 01   00:00 BST (GMT + 1hr)

April 15   23:00 BST (GMT + 1hr)

April – Looking South

Leo is the most prominent constellation in the southern sky in April, and vaguely looks like the creature after which it is named. **Gemini**, with **Castor** and **Pollux**, remains clearly visible in the west, and **Cancer** lies between the two constellations. To the east of Leo, the whole of **Virgo**, with **Spica** (α Virginis) its brightest star, is well clear of the horizon. Below Leo and Virgo, the complete length of **Hydra** is visible, running below Leo and Virgo, with **Alphard** (α Hydrae) half-way between Regulus and the southwestern horizon. Farther east, the two small constellations of **Crater** and the rather brighter **Corvus** lie between Hydra and Virgo.

Boötes and **Arcturus** are prominent in the eastern sky, together with the circlet of **Corona Borealis**, which is framed by Boötes and the neighbouring constellation of **Hercules**. Between Leo and Boötes lies the tiny constellation of **Coma Berenices**, most notable for being the location of the Coma Cluster of galaxies. There are about 1000 galaxies in this cluster, which is located near the North Galactic Pole, where we are looking out of the plane of the Galaxy and are thus able to see deep into space. Only about ten of the brightest galaxies in the Coma Cluster are visible with the largest amateur telescopes.

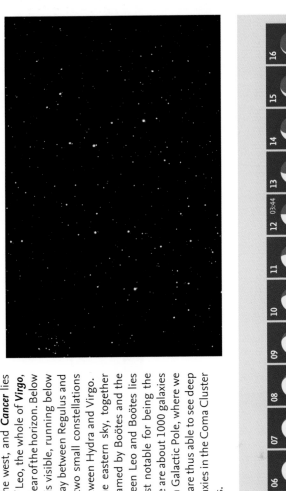

The highly distinctive constellation of Leo, with first-magnitude Regulus on the west, and vaguely looks like the creature after the 'Sickle' of stars above, and stretching east to the second-brightest star, Denebola.

The Moon's phases for April

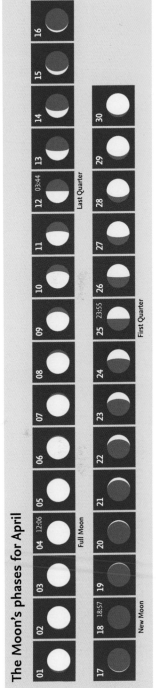

01 | 02 | 03 | 04 12:06 | 05 | 06 | 07 | 08 | 09 | 10 | 11 | 12 03:44 | 13 | 14 | 15 | 16

Full Moon — Last Quarter

17 18:57 | 18 | 19 | 20 | 21 | 22 | 23 | 24 | 25 23:55 | 26 | 27 | 28 | 29 | 30

New Moon — First Quarter

April – Moon and Planets

The Moon

Full Moon is on April 4, when there is a total lunar eclipse at 12:01 UT, but this is visible only from the Pacific and eastern Australia. On April 5, as the Moon sets in the west, it passes just north of *Spica* (α Virginis). On April 8–9, the waning gibbous Moon passes close to both *Saturn* and *Antares* (α Scorpii). On April 18, just before New Moon, it occults Uranus as seen from the western Pacific and Indonesia. On April 21, the waxing crescent passes close and north of *Aldebaran* (α Tauri) and then on April 25–26, south of *Jupiter* (in *Cancer*) and *Regulus* (α Leonis). The minor planet (3)*Juno*, at magnitude 10.0 in *Cancer*, is occulted on April 26 at 07:10 UT, but this is invisible from Europe, and seen only from a wide swathe of the central Pacific Ocean.

The Planets

Mercury is initially too close to the Sun to be visible, but moves towards eastern elongation (on May 7) at the end of the month. *Venus* is visible early in the night, brightening slightly from -4.0 to -4.1 by May. It sets about 22:00 at the start of the month and 23:00 by the end as it moves from *Aries* to *Taurus*. It is close to *Aldebaran* (α Tauri) on April 21. *Mars* may be glimpsed in the west early in the month but soon becomes lost in evening twilight. *Jupiter* is in Cancer, easily visible for most of the night, setting about 02:00 early in the month and about 04:00 at the end. It brightens slightly from magnitude -2.3 to -2.1 over the month. *Saturn*, in *Scorpius*, rises about midnight (UT) at the beginning of the month, and is slightly higher at the end of April. It brightens slightly over the month, from magnitude 0.3 to 0.1.

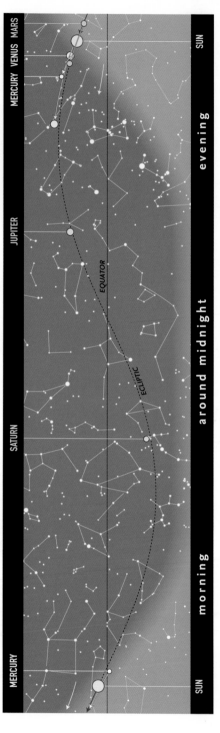

The path of the Sun and the planets along the ecliptic in April.

Calendar for April

01	13:01	Moon at apogee
04	12:01	Total lunar eclipse (Pacific, E. Australia)
04	12:06	Full Moon
05	04:47	Spica 3.5°S of Moon
08	12:45	Saturn 2.2°S of Moon
08	21:19	Antares 9.2°S of Moon
12	03:44	Last Quarter
16–25		April Lyrid meteor shower
17	03:48	Moon at perigee
18	00:34	Uranus 0.0°N of Moon (occultation over Pacific, Indonesia, 00:34 UT – in daylight)
18	18:57	New Moon
Apr.19–May 26		Eta Aquarid meteor shower
19	11:05	Mercury 3.5°N of Moon
19	19:02	Mars 3.1°N of Moon
21	04:00 *	Aldebaran 7.5°S of Venus
21	16:57	Aldebaran 0.9°S of Moon
21	18:08	Venus 6.6°N of Moon
21–22		April Lyrid shower maximum
25	04:03	Regulus 4.0°N of Moon
25	23:55	First Quarter
26	18:12	Jupiter 5.5°N of Moon
29	03 55	Moon at apogee

* These objects are close together for an extended period around this time.

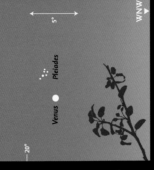

Evening 21:30 (BST)

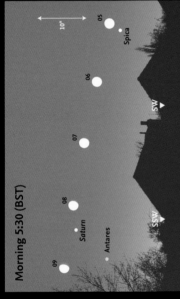

Morning 5:30 (BST)

April 5–9 • The Moon passes Spica, Saturn and Antares.

April 11 • Venus close to the Pleiades, in the evening sky.

Evening 22:30 (BST)

Evening 21:30 (BST)

April 21 • The Moon with Aldebaran, Venus and the Pleiades, after sunset.

April 25–28 • The Moon passes Jupiter and Regulus.

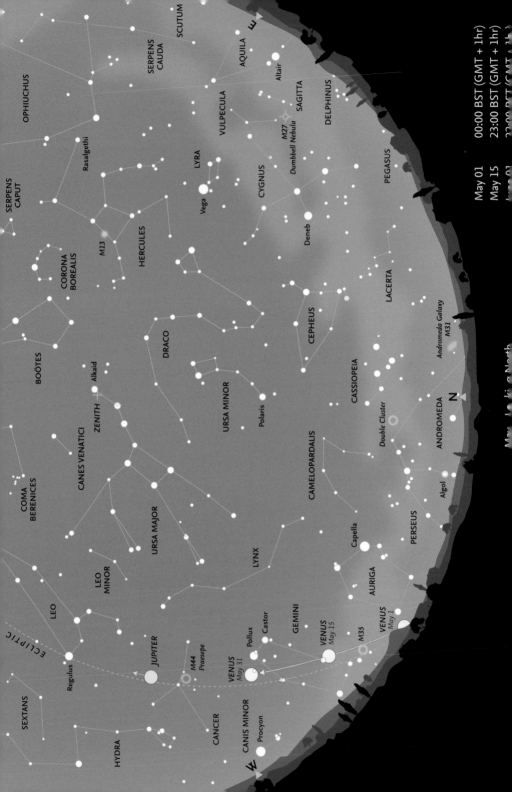

SCUTUM

SERPENS
CAUDA

OPHIUCHUS

AQUILA

Altair

SERPENS
CAPUT

Rasalgethi

VULPECULA

M27
Dumbbell Nebula

SAGITTA

DELPHINUS

M13

LYRA

HERCULES

Vega

CYGNUS

PEGASUS

CORONA
BOREALIS

Deneb

CEPHEUS

LACERTA

BOÖTES

DRACO

Andromeda Galaxy
M31

ZENITH

Alkaid

URSA MINOR

CASSIOPEIA

CANES VENATICI

Polaris

Double Cluster

N

COMA
BERENICES

CAMELOPARDALIS

ANDROMEDA

Algol

PERSEUS

URSA MAJOR

LEO
MINOR

LYNX

Capella

LEO

AURIGA

ECLIPTIC

JUPITER

Pollux

Castor

GEMINI

VENUS
May 15

M35

VENUS
May 1

Regulus

M44
Praesepe

SEXTANS

VENUS
May 31

CANCER

CANIS MINOR

Procyon

HYDRA

W

May looking North

May – Looking North

Cassiopeia is now low over the northern horizon and, to its west, the southern portions of both **Perseus** and **Auriga** are becoming difficult to observe. **Gemini**, with **Castor** and **Pollux**, is sinking towards the western horizon.

In the east, two of the stars of the 'Summer Triangle', **Vega** in **Lyra** and **Deneb** in **Cygnus**, are clearly visible, and the third, **Altair** in **Aquila**, is beginning to climb above the horizon. The sprawling constellation of **Hercules** is high in the east and the brightest globular cluster in the northern hemisphere, M13, is visible to the naked eye on the western side of the 'Keystone'.

Later in the night (and in the month) the westernmost stars of **Pegasus** begin to come into view, while the stars of **Andromeda** are skimming the northeastern horizon. High overhead, **Alkaid** (η Ursae Majoris), the last star in the 'tail' of the Great Bear, is close to the zenith, while the main body of the constellation has swung round into the western sky.

Meteors

The **Eta Aquarids** are one of the two meteor showers associated with Comet 1P/Halley (the other being the **Orionids**, in October). The Eta Aquarids are not particularly favourably placed for northern-hemisphere observers, because the radiant is near the celestial equator, near the 'Water Jar' in **Aquarius**, well below the horizon until late in the night (around dawn). There is a radiant map for the Eta Aquarids on page 16.

Their maximum in 2015, on May 5–6, occurs a day after Full Moon, so there will be great interference from moonlight. A large proportion (about 25 per cent) of the meteors leave persistent trains.

Cygnus, sometimes known as the 'Northern Cross', depicts a swan flying down the Milky Way towards Sagittarius. The brightest star, Deneb (α Cygni), represents the tail, and Albireo (β Cygni) marks the position of the head, and may be found in the lower right section of the image.

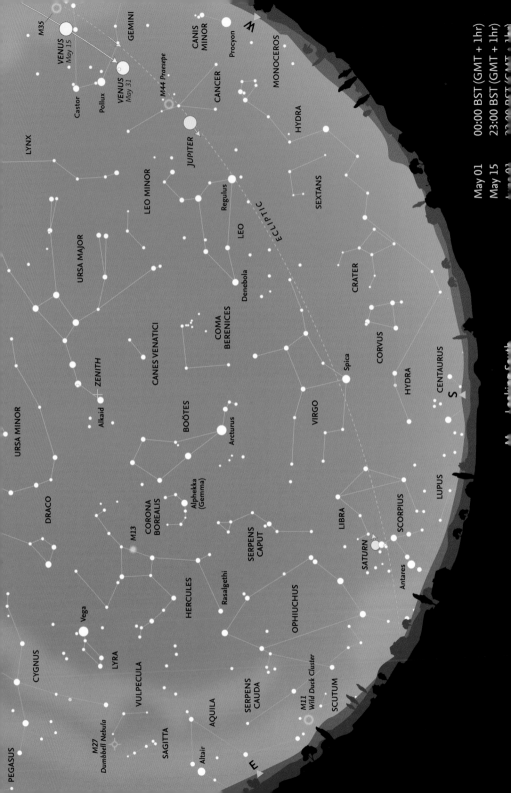

M35

VENUS
May 15

VENUS
May 31

M44 Praesepe

Castor

Pollux

GEMINI

CANIS
MINOR

Procyon

MONOCEROS

CANCER

LYNX

JUPITER

LEO MINOR

HYDRA

Regulus

LEO

SEXTANS

ECLIPTIC

URSA MAJOR

Denebola

CRATER

ZENITH

Alkaid

CANES VENATICI

COMA
BERENICES

CORVUS

URSA MINOR

Spica

HYDRA

CENTAURUS

DRACO

BOÖTES

VIRGO

S

Arcturus

Alphekka
(Gemma)

CORONA
BOREALIS

LIBRA

LUPUS

M13

SERPENS
CAPUT

SCORPIUS

CYGNUS

Vega

HERCULES

Rasalgethi

OPHIUCHUS

SATURN

Antares

LYRA

VULPECULA

M27
Dumbbell Nebula

SERPENS
CAUDA

M11
Wild Duck Cluster

SCUTUM

AQUILA

SAGITTA

Altair

PEGASUS

E

May – Looking South

Early in the night, the constellation of *Virgo*, with *Spica* (α Virginis), lies due south, with *Leo* and both *Regulus* and *Denebola* (α and β Leonis, respectively) to its west still well clear of the horizon. Later in the night, the rather faint zodiacal constellation of *Libra* becomes visible and, to its east, the ruddy star *Antares* (α Scorpii) begins to climb up over the horizon.

Arcturus in *Boötes* is high in the south, with the distinctive circlet of *Corona Borealis* clearly visible to its east. The brightest star (α Coronae Borealis) is known as *Alphekka* or *Gemma*. The large constellation of *Ophiuchus* (which actually crosses the ecliptic, and is thus the 'thirteenth' zodiacal constellation) is climbing into the eastern sky. Before the constellation boundaries were formally adopted by the International Astronomical Union in 1930, the southern region of Ophiuchus was regarded as forming part of the constellation of Scorpius, which had been part of the Zodiac since antiquity.

The constellation of Virgo. The brightest object is the planet Mars, which is again in this constellation in November and December 2015.

The Moon's phases for May

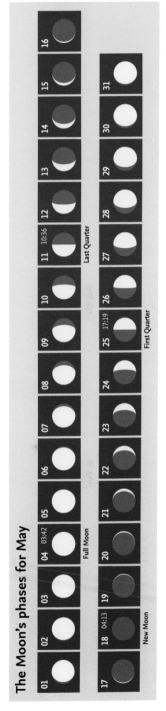

May – Moon and Planets

The Moon

On May 1–2, the waxing gibbous Moon passes north of first-magnitude *Spica* (α Virginis). On May 5 (one day after Full) it passes north of *Saturn* (in *Scorpius* between *Libra* and *Ophiuchus*) and later the next day well north of *Antares* (α Scorpii). It passes close to both *Mars* and *Aldebaran* in the middle of the month, but these events occur in daylight. On May 21, the Moon passes south of *Venus* shortly after sunset. On May 24 it is close to *Jupiter* (in *Cancer*) in the western sky, and the next day passes north of *Regulus*. Towards the end of the month (May 29) it again passes north of Spica.

The Planets

Mercury reaches greatest eastern elongation on May 7 at magnitude 0.2. *Venus* fades slightly during the month, from magnitude -4.1 to -4.3, and is in *Gemini*; setting around midnight. Early in the month, *Mars* may be glimpsed (at magnitude 1.4–1.5) low in the west just before it sets. *Jupiter* remains in *Cancer*, setting about 02:00 early in the month and around 00:00 at the end. *Saturn* is retrograding, moving from *Scorpius* into *Libra* during the month, where it comes to opposition at magnitude 0.0 on May 23.

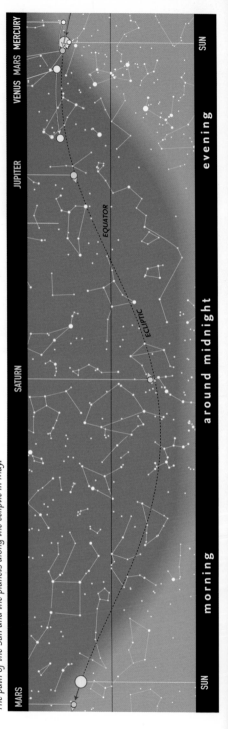

The path of the Sun and the planets along the ecliptic in May.

Calendar for May

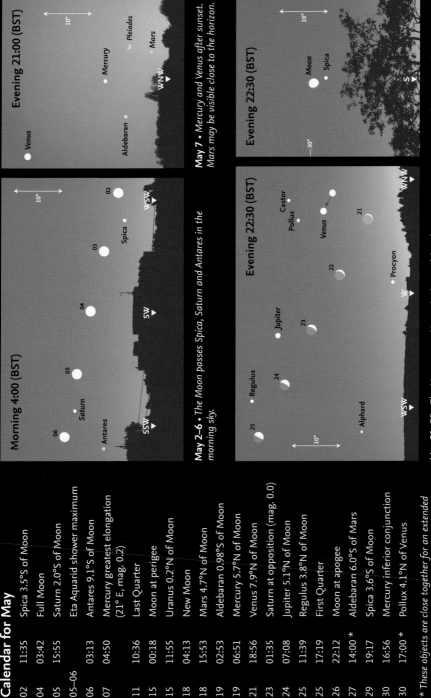

02	11:35	Spica 3.5°S of Moon
04	03:42	Full Moon
05	15:55	Saturn 2.0°S of Moon
05–06		Eta Aquarid shower maximum
06	03:13	Antares 9.1°S of Moon
07	04:50	Mercury greatest elongation (21° E, mag. 0.2)
11	10:36	Last Quarter
15	00:18	Moon at perigee
15	11:55	Uranus 0.2°N of Moon
18	04:13	New Moon
18	15:53	Mars 4.7°N of Moon
19	02:53	Aldebaran 0.98°S of Moon
19	06:51	Mercury 5.7°N of Moon
21	18:56	Venus 7.9°N of Moon
23	01:35	Saturn at opposition (mag. 0.0)
24	07:08	Jupiter 5.1°N of Moon
25	11:39	Regulus 3.8°N of Moon
25	17:19	First Quarter
26	22:12	Moon at apogee
27	14:00 *	Aldebaran 6.0°S of Mars
29	19:17	Spica 3.6°S of Moon
30	16:56	Mercury inferior conjunction
30	17:00 *	Pollux 4.1°N of Venus

* These objects are close together for an extended period around this time.

Morning 4:00 (BST)

Venus · 06 · Saturn · 05 · 04 · Antares · SSW · 03 · Spica · 02 · SW · WSW · 10°

May 2–6 • The Moon passes Spica, Saturn and Antares in the morning sky.

Evening 21:00 (BST)

Venus · Mercury · ·.· Pleiades · Mars · Aldebaran · WNW · 10°

May 7 • Mercury and Venus after sunset. Mars may be visible close to the horizon.

Evening 22:30 (BST)

Regulus · 25 · 24 · 23 · Jupiter · 22 · Venus · Pollux · Castor · 21 · Procyon · Alphard · WSW · W · WNW · 10°

May 21–25 • The Moon passes Venus, Jupiter and Regulus.

Evening 22:30 (BST)

Moon · Spica · –30° · S · 10°

May 29 • The Moon and Spica close together.

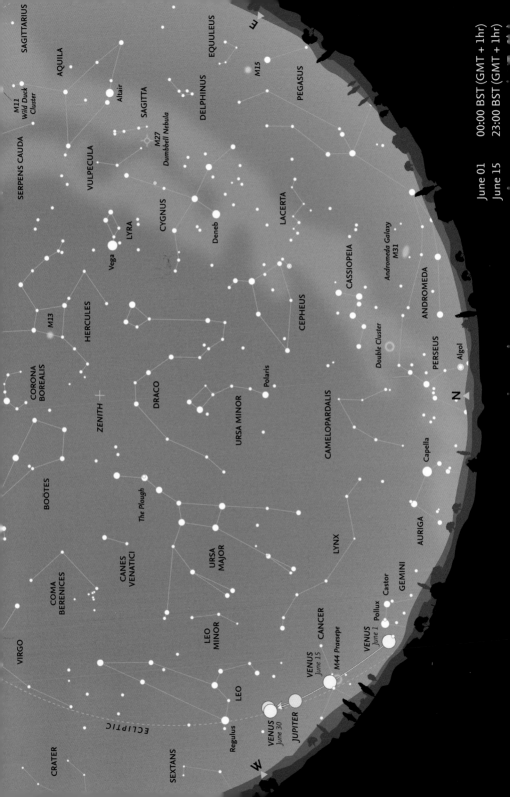

SAGITTARIUS

AQUILA

M11
Wild Duck
Cluster

Altair

SERPENS CAUDA

SAGITTA

EQUULEUS

M15

DELPHINUS

PEGASUS

M27
Dumbbell Nebula

VULPECULA

LACERTA

CYGNUS

Deneb

LYRA

Vega

CASSIOPEIA

CEPHEUS

Andromeda Galaxy
M31

HERCULES

M13

ANDROMEDA

CORONA
BOREALIS

DRACO

Double Cluster

PERSEUS

Polaris

Algol

ZENITH

URSA MINOR

N

CAMELOPARDALIS

BOÖTES

The Plough

Capella

CANES
VENATICI

URSA
MAJOR

LYNX

AURIGA

COMA
BERENICES

LEO
MINOR

GEMINI

VIRGO

Pollux
Castor

LEO

CANCER

M44 Praesepe

VENUS
June 1

VENUS
June 15

CRATER

SEXTANS

Regulus

VENUS
June 30

JUPITER

ECLIPTIC

W

E

June 01   00:00 BST (GMT + 1hr)
June 15   23:00 BST (GMT + 1hr)

June – Looking North

As we approach the summer solstice (June 21), even in southern England and Ireland a form of twilight persists throughout the night and, farther north, in Scotland, the sky remains so light that most of the fainter stars and constellations are invisible. Even brighter stars, such as the seven stars making up the well-known asterism known as the *Plough* in *Ursa Major* may be difficult to detect except around local midnight, 00:00 UT (01:00 BST).

But there is one compensation during these light nights: even southern observers may be lucky enough to witness a display of noctilucent clouds (NLC). These are highly distinctive clouds shining with an electric-blue tint, observed in the sky in the direction of the North Pole. They are the highest clouds in the atmosphere, occurring at altitudes of 80–85 km, far above all other clouds. They are only visible during summer nights, for about a month or six weeks on either side of the solstice, when observers are in darkness, but the clouds themselves remain illuminated by sunlight, reaching them from the Sun, itself hidden below the northern horizon.

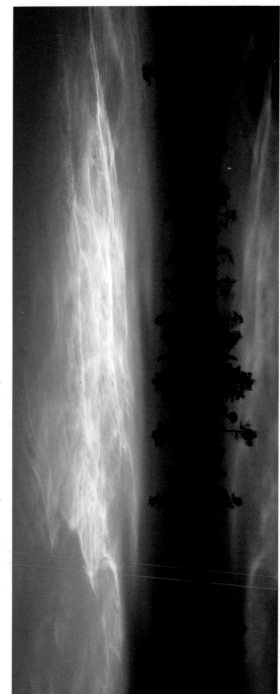

Noctilucent clouds consist of ice particles, formed on tiny specks of meteoric dust from space or on clusters of ions (charged particles), created by cosmic rays.

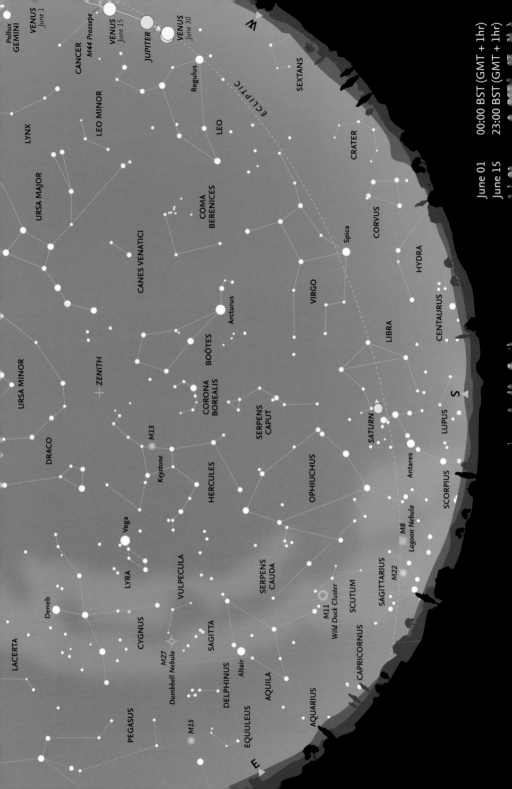

June – Looking South

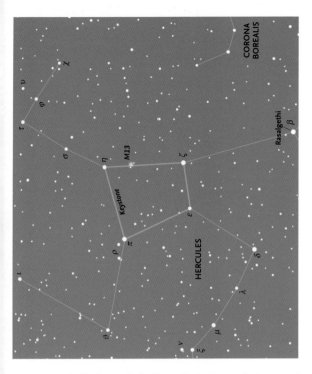

Although the persistent twilight makes observing even the southern sky difficult, the rather undistinguished constellation of **Libra** lies almost due south. The red supergiant star **Antares** – the name means the 'Rival of Mars' – in **Scorpius** is visible slightly to the east of the meridian, but the 'tail' or 'sting' remains below the horizon. Higher in the sky is the large constellation of **Ophiuchus** (the 'Serpent Bearer'), lying between the two halves of the constellation of **Serpens**: **Serpens Caput** ('Head of the Serpent') to the west and **Serpens Cauda** ('Tail of the Serpent') to the east. (Serpens is the only constellation to be divided into two distinct parts.) The ecliptic runs across Ophiuchus, and the Sun actually spends far more time in the constellation than it does in the 'classical' zodiacal constellation of Scorpius, a small area of which lies between Libra and Ophiuchus.

Higher in the southern sky, the three constellations of **Boötes**, **Corona Borealis** and **Hercules** are now better placed for observation than at any other time of the year.

Finder chart for M13 in Hercules. All stars brighter than mag. 7.5 are shown.

The Moon's phases for June

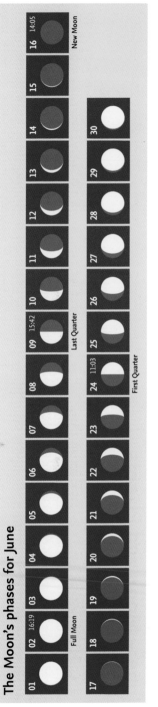

June – Moon and Planets

The Moon

Perigee occurs on June 10, and this is the most distant of the year, at 369,711 km. The next day, June 11, the Moon occults *Uranus*, but the event is visible only from Australia. An occultation of *Mercury* occurs on June 15, but the event occurs in daylight over the Pacific. On June 20, the Moon passes south of *Venus* and *Jupiter* and, the next day, *Regulus* (α Leonis). On June 29, the Moon passes just north of Saturn shortly before the planet sets.

The Planets

On June 24, *Mercury* (magnitude 0.4) comes to greatest western elongation, but is too low in the dawn sky to be readily visible. For the whole of June, *Venus* and *Jupiter* (in *Cancer* and *Leo*, respectively) are close to one another in the evening sky, and visible for about an hour before setting in the west-northwest. *Venus* is at eastern elongation on June 6, when it is very bright at magnitude -4.4. *Mars* is too close to the Sun to be visible, being at superior conjunction on June 14. *Saturn* (magnitude 0.1–0.2) is in *Libra* and visible throughout the few hours of darkness. On June 11, *Uranus*, just visible before sunrise from Europe, is actually occulted by the Moon for observers in Australia.

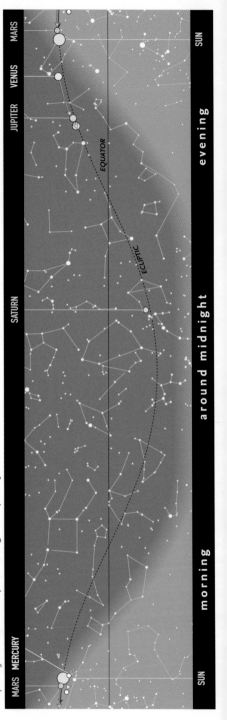

The path of the Sun and the planets along the ecliptic in June.

01	19:38	Saturn 1.9°S of Moon
02	10:28	Antares 9.1°S of Moon
02	16:19	Full Moon
06	18:29	Venus greatest elongation (45° E, mag. -4.4)
09	15:42	Last Quarter
10	04:44	Moon at perigee
11	20:26	Uranus 0.6°N of Moon
13	08:59	Venus 0.6°N of Praesepe
14	15:56	Mars at superior conjunction
15	02:26	Mercury 0.0°N of Moon
15	11:33	Aldebaran 1.0°S of Moon
16	12:40	Mars 5.5°N of Moon
16	14:05	New Moon
20	11:28	Venus 5.8°N of Moon
20	23:34	Jupiter 4.7°N of Moon
21	16:38	Summer solstice
21	19:34	Regulus 3.6°N of Moon
23	17:00	Moon at apogee
24	11:03	First Quarter
24	17:08	Mercury greatest elongation (23° W, mag. 0.4)
26	03:23	Spica 3.8°S of Moon
29	01:03	Saturn 2.0°S of Moon
29	19:07	Antares 9.2°S of Moon

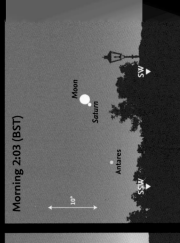

Evening 23:00 (BST)

Saturn · Antares · SSE

Evening 23:00 (BST)

Regulus · Jupiter · Venus · Pollux · Castor · WNW · W

June 1–2 • *The Moon passes Saturn and Antares.*

June 6 • *Venus has reached greatest eastern elongation. Jupiter nearby.*

Morning 2:03 (BST)

Moon · Saturn · Antares · SW · SSW

Evening 22:30 (BST)

Algieba · Regulus · Jupiter · Venus · WNW · W

June 19–21 • *The Moon passes Venus, Jupiter and Regulus.*

June 29 • *The Moon and Saturn in conjunction.*

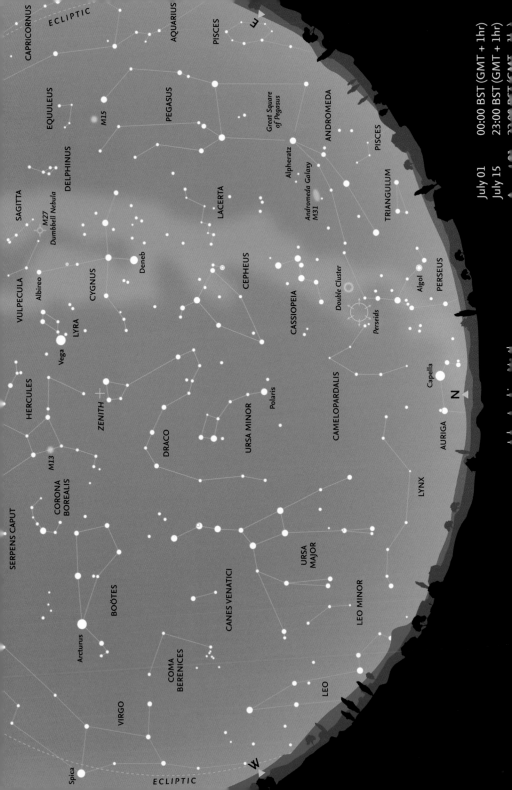

CAPRICORNUS
ECLIPTIC
AQUARIUS
PISCES
E

EQUULEUS
M15
PEGASUS
Great Square
of Pegasus
ANDROMEDA
PISCES
DELPHINUS
Alpheratz
Andromeda Galaxy
M31
TRIANGULUM
LACERTA
SAGITTA
M27
Dumbbell Nebula
Deneb
CYGNUS
CEPHEUS
Double Cluster
Algol
PERSEUS
VULPECULA
Albireo
CASSIOPEIA
Perseids
LYRA
Vega
HERCULES
ZENITH
Capella
CAMELOPARDALIS
N
M13
URSA MINOR
AURIGA
CORONA
BOREALIS
Polaris
DRACO
SERPENS CAPUT
LYNX
BOÖTES
URSA
MAJOR
CANES VENATICI
Arcturus
LEO MINOR
COMA
BERENICES
LEO
VIRGO
W
Spica
ECLIPTIC

July 01   00:00 BST (GMT + 1hr)
July 15   23:00 BST (GMT + 1hr)

July – Looking North

As in June, light nights and the chance of observing noctilucent clouds persist throughout July, but later in the month (and particularly after midnight) some of the major constellations begin to be more easily seen. **Capella**, the brightest star in **Auriga** (most of which is too low to be visible), is skimming the northern horizon. **Cassiopeia** is clearly visible in the northeast and **Perseus**, to its south, is beginning to climb clear of the horizon. The band of the Milky Way, from Perseus through Cassiopeia towards **Cygnus**, stretches up into the northeastern sky. If the sky is dark and clear, you may be able to make out the small, faint constellation of **Lacerta**, lying across the Milky Way between Cassiopeia and Cygnus. In the east, the stars of **Pegasus** are now well clear of the horizon, with the main line of stars forming **Andromeda** roughly parallel to the horizon in the northeast. **Alpheratz** (α Andromedae) is actually the star at the north-eastern corner of the Great Square of **Pegasus**. **Cepheus** and **Ursa Major** are on opposite sides of **Polaris** and **Ursa Minor**, in the east and west, respectively. The head of **Draco** is very close to the zenith so the whole of this winding constellation is readily seen.

Meteors

July brings increasing meteor activity, mainly because there are several minor radiants active in the constellations of **Capricornus** and **Aquarius.** Because of their location, however, observing conditions are not particularly favourable for northern-hemisphere observers, although the first shower, the **Alpha Capricornids**, active from July 11 to August 10 (peaking July 28–29), does often produce very bright fireballs. The maximum rate, however, is only about five per hour. The parent body is Comet 169P/NEAT. The most prominent shower is probably that of the **Delta Aquarids**, which are active from around July 21 to August 23, with a peak on July 27–28, although even then the hourly rate is

unlikely to reach 20 meteors per hour. In this case, the parent body is possibly Comet 96P/Machholz. This year, unfortunately, both shower maxima occur just before Full Moon, so observational conditions are unfavourable. The **Perseids**, by contrast, begin on July 13 and peak on August 11–12, three days before New Moon. A chart showing the **Delta Aquarid** radiant is shown on p.16.

In July 2015, the New Horizons spaceprobe, launched by NASA in 2006, will finally arrive at Pluto, the dwarf planet, now known to be one of several similar bodies, or Trans-Neptunian Objects (TNOs), orbiting in the distant Solar System. It will be our only chance to study Pluto and its major satellite, Charon, for many years to come.

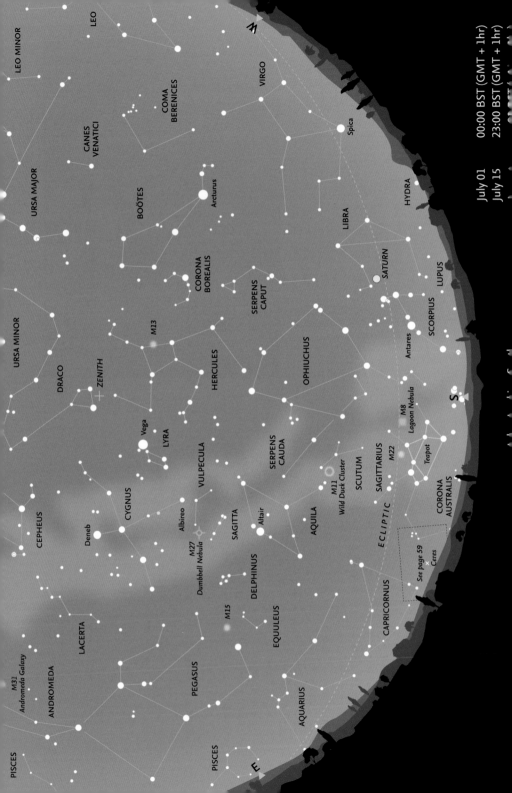

LEO MINOR

LEO

LEO MINOR

URSA MAJOR

CANES
VENATICI

COMA
BERENICES

VIRGO

Spica

BOÖTES

Arcturus

HYDRA

LIBRA

URSA MINOR

CORONA
BOREALIS

M13

SERPENS
CAPUT

SATURN

LUPUS

DRACO

HERCULES

OPHIUCHUS

SCORPIUS

ZENITH

Antares

Vega

LYRA

SERPENS
CAUDA

M8
Lagoon Nebula

S

CYGNUS

VULPECULA

M22

Teapot

SAGITTARIUS

Deneb

Albireo

SAGITTA

Altair

SCUTUM

ECLIPTIC

CORONA
AUSTRALIS

M27
Dumbbell Nebula

AQUILA

M11
Wild Duck Cluster

CEPHEUS

DELPHINUS

See page 59

Ceres

LACERTA

M15

EQUULEUS

CAPRICORNUS

M31
Andromeda Galaxy

PEGASUS

ANDROMEDA

AQUARIUS

PISCES

PISCES

E

July – Looking South

Although part of the constellation remains hidden, this is perhaps the best time of year to see **Scorpius**, with deep red **Antares** (α Scorpii), glowing just above the southern horizon. At around midnight (UT), 01:00 BST, part of **Sagittarius**, with the distinctive asterism of the 'Teapot', and the dense star clouds of the centre of the Milky Way, are just visible in the south. With clear skies, the Great Rift – actually dust clouds that hide the more distant stars – runs down the Milky Way from Cygnus towards Sagittarius. The sprawling constellation of **Ophiuchus** lies close to the meridian for a large part of the month, separating the two halves of the constellation of **Serpens**. The western half is called **Serpens Caput** (Head of the Serpent) and the eastern part **Serpens Cauda** (Tail of the Serpent). In the east, the bright **Summer Triangle**, consisting of **Vega** in **Lyra**, **Deneb** in **Cygnus** and **Altair** in **Aquila**, begins to dominate the southern sky, as it will throughout August and into September. The small constellation of Lyra, with Vega and a distinctive quadrilateral of stars to its east and south, lies not far south of the zenith.

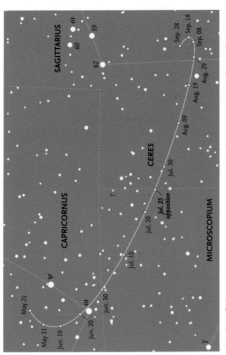

The dwarf planet Ceres comes to opposition on July 25 at magnitude 7.5, but is too low to be readily observed. The chart shows all stars brighter than magnitude 8.6.

The Moon's phases for July

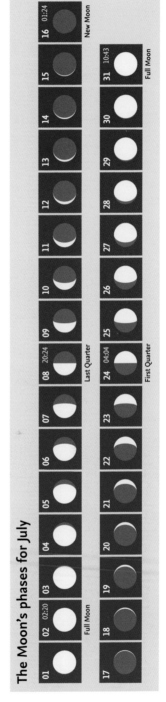

July – Moon and Planets

The Moon

Full Moon occurs twice in July, on July 2 (in **Sagittarius**) and July 31 (in **Capricornus**). On July 9, the waning crescent passes south of **Uranus** (at magnitude 5.8) in **Pisces**. On July 12, the Moon is close to **Aldebaran** (α Tauri) in the brightening dawn sky. Later in the month, on July 23, it passes close to **Spica** (α Virginis). Closest approach is at 11:02 in daylight, but the two objects will be visible near one another for a couple of hours before Spica sets.

The Planets

In July, **Venus** and **Jupiter** are visible near one another in the evening twilight. Closest approach actually occurs on July 1 in daylight. **Mercury** reaches superior conjunction (on the far side of the Sun) on July 23, so is invisible throughout the month. On July 6, the **Earth** is at aphelion, farthest from the Sun. **Mars** is near Mercury in the sky and again is too close to the Sun to be seen. **Saturn** (in Libra) is visible early in the month towards the southwest and setting just before midnight, becoming more difficult at the end of July, setting just before midnight (UT). The dwarf planet **(1) Ceres** comes to opposition on July 25 at magnitude 7.5, but is on the border of Sagittarius and Microscopium, and too low to be readily observed (see chart on page 58).

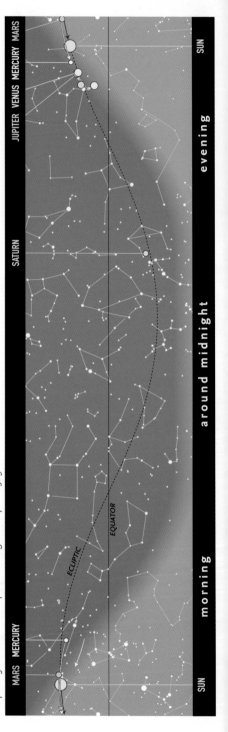

The path of the Sun and the planets along the ecliptic in July.

Calendar for July

01	14:00 ★	Jupiter 0.4°N of Venus
02	02:20	Full Moon
05	18:52	Moon at perigee
06		Earth at aphelion (1.016682122 AU)
08	20:24	Last Quarter
09	02:47	Uranus 0.8°N of Moon
Jul.11–Aug.10		Alpha Capricornid meteor shower
12	18:18	Aldebaran 0.9°S of Moon
Jul.13–Aug.26		Perseid meteor shower
16	01:24	New Moon
21	11:02	Moon at apogee
Jul.21–Aug.23		Delta Aquarid meteor shower
23	11:11	Spica 4.1°S of Moon
23	19:24	Mercury at superior conjunction
24	04:04	First Quarter
25	07:54	Ceres opposition (mag. 7.5)
26	08:19	Saturn 2.2°S of Moon
27	04:21	Antares 9.3°S of Moon
27–28		Delta Aquarid shower maximum
28–29		Alpha Capricornid shower maximum
31	10:43	Full Moon

★ These objects are close together for an extended period around this time.

Evening 22:30 (BST)

July 1 • Venus and Jupiter close together in the evening sky. Regulus, to the east (left), is much fainter.

Morning 4:00 (BST)

July 13 • The crescent Moon and Aldebaran, before sunrise.

Morning 22:00 (BST)

July 22–27 • The Moon passes the almost equally bright Spica (mag. 0.98), Saturn (mag. 1.00) and Antares (mag. 1.06).

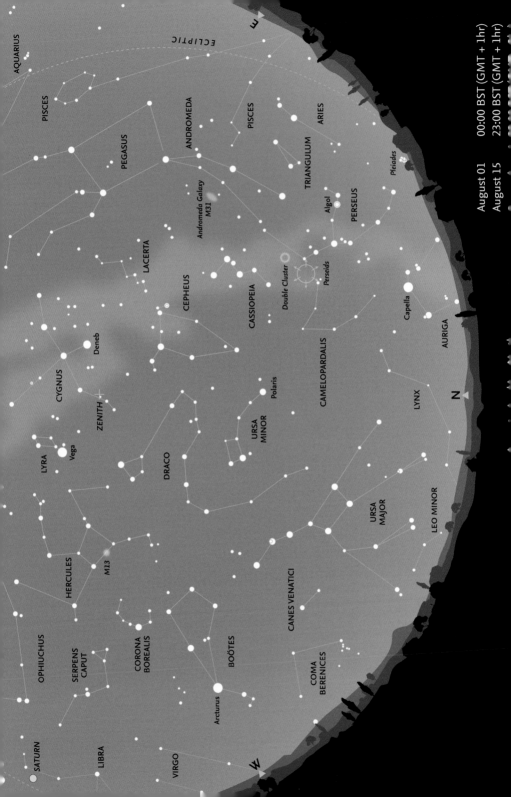

AQUARIUS

PISCES

PEGASUS

ANDROMEDA

PISCES

ARIES

Pleiades

Andromeda Galaxy
M31

TRIANGULUM

PERSEUS

Algol

LACERTA

Perseids

Double Cluster

CEPHEUS

CASSIOPEIA

Capella

CAMELOPARDALIS

AURIGA

ECLIPTIC

E

Deneb

CYGNUS

ZENITH +

Polaris

LYNX

N

URSA
MINOR

LYRA

Vega

DRACO

URSA
MAJOR

LEO MINOR

HERCULES

M13

CANES VENATICI

OPHIUCHUS

SERPENS
CAPUT

CORONA
BOREALIS

BOÖTES

COMA
BERENICES

SATURN

LIBRA

VIRGO

Arcturus

W

August – Looking North

Ursa Major is now the 'right way up' in the northwest, although some of the fainter stars in the south of the constellation are difficult to see. Beyond it, **Boötes** stands almost vertically in the west, but pale orange **Arcturus** is sinking towards the horizon. Higher in the sky, both **Corona Borealis** and **Hercules** are clearly visible.

In the northeast, **Capella** is clearly visible, but most of **Auriga** still remains below the horizon. Higher in the sky, **Perseus** is gradually coming into full view and, later in the night and later in the month, the beautiful **Pleiades** cluster rises above the northeastern horizon. Between Perseus and **Polaris** lies the faint and unremarkable constellation of **Camelopardalis**.

Higher still, both **Cassiopeia** and **Cepheus** are well placed for observation, despite the fact that Cassiopeia is completely immersed in the band of the Milky Way, as is the 'base' of Cepheus. **Pegasus** and **Andromeda** are now well above the eastern horizon and, below them, the constellation of **Pisces** is climbing into view. Two of the stars in the **Summer Triangle**, **Deneb** and **Vega**, are close to the zenith high overhead.

Meteors

August is regarded by astronomers as the month when one of the best meteor showers of the year occurs: the **Perseids**. This is a long shower, generally beginning about July 13 and continuing until around August 26, with a maximum on August 11–12, when the rate may reach as high as 80 meteors per hour (and on rare occasions, even higher). In 2015, maximum is three days before New Moon, so interference by moonlight will not be a major factor. The Perseids are debris from Comet 109P/Swift-Tuttle (the Great Comet of 1862). Perseid meteors are fast and many of the brighter ones leave persistent trains.

The constellation of Perseus is not only the location of the radiant for the Perseid meteor shower, but is also well-known for the pair of open clusters, known as the Double Cluster (near the top edge of the image) close to the border with Cassiopeia, and also for Algol (β Persei), a famous variable star that may be found at the centre of the photograph. The open star cluster, the Pleiades, in Taurus, is close to the bottom edge of the image.

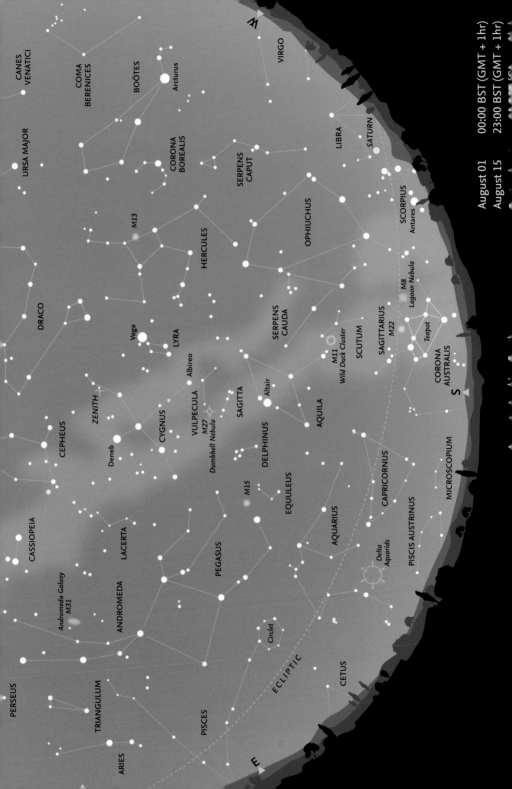

N

VIRGO

CANES
VENATICI

COMA
BERENICES

BOÖTES

Arcturus

URSA MAJOR

CORONA
BOREALIS

SERPENS
CAPUT

LIBRA

SATURN

M13

HERCULES

OPHIUCHUS

SCORPIUS

Antares

DRACO

SERPENS
CAUDA

M8
Lagoon Nebula

Vega

LYRA

SCUTUM

SAGITTARIUS

M22

M11
Wild Duck Cluster

Teapot

CORONA
AUSTRALIS

S

Albireo

SAGITTA

Altair

ZENITH

CYGNUS

VULPECULA

M27
Dumbbell Nebula

DELPHINUS

AQUILA

CEPHEUS

Deneb

M15

EQUULEUS

MICROSCOPIUM

CASSIOPEIA

LACERTA

PEGASUS

AQUARIUS

CAPRICORNUS

PISCIS AUSTRINUS

Andromeda Galaxy
M31

ANDROMEDA

Delta
Aquarids

PERSEUS

TRIANGULUM

PISCES

Circlet

ECLIPTIC

CETUS

ARIES

E

August 01   00:00 BST (GMT + 1hr)
August 15   23:00 BST (GMT + 1hr)

August – Looking South

The whole stretch of the summer Milky Way stretches across the sky in the south, from **Cygnus**, high in the sky near the zenith, past **Aquila**, with bright **Altair** (α Aquilae), to part of the constellation of **Sagittarius** close to the horizon, where the pattern of stars known as the 'Teapot' may be just visible. Between **Albireo** (β Cygni) and Altair lie the two small constellations of **Vulpecula** and **Sagitta**, with the latter easier to distinguish (because of its shape) from the clouds of the Milky Way. Between Sagitta and **Pegasus** to the east lie the highly distinctive five stars that form the tiny constellation of **Delphinus** (again, one of the few constellations that actually bear some resemblance to the creatures after which they are named). Below Aquila, mainly in the star clouds of the Milky Way, lies **Scutum**, most famous for the bright open cluster, M11 or the 'Wild Duck Cluster', readily visible in binoculars. To the southeast of Aquila lie the two zodiacal constellations of **Capricornus** and **Aquarius**.

Aquila, with Altair (α Aquilae) its brightest star, like Cygnus, represents a bird – in this case an eagle – flying down the length of the Milky Way.

The Moon's phases for August

01	02	03	04	05	06	07 02:03	08	09	10	11	12	13	14 14:53	15	16

Last Quarter

New Moon

17	18	19	20	21	22 19:31	23	24	25	26	27	28	29 18:35	30	31

First Quarter

Full Moon

66

August – Moon and Planets

The Moon

The Moon reaches perigee twice in August, on August 2 and 30. On August 8, as *Taurus* rises in the east, the Moon passes close to *Aldebaran* (α Tauri). On August 22, the First Quarter Moon lies north of *Saturn* (in *Libra*). Both bodies are visible for a short period before setting in the northwest. The Moon then passes due north of *Antares* (α Scorpii) the next day, August 23, when the star is also visible for a while in the evening sky just before it sets.

The Planets

Mercury and *Jupiter* are east of the Sun and too close to it to be observed. *Venus*, initially to the east, rapidly passes through inferior conjunction (between the Sun and Earth) on August 15 and becomes visible before dawn towards the end of the month, brightening to magnitude -4.5. *Mars* is also to the west of the Sun, in the same constellation as Venus (Cancer), but much fainter at magnitude 1.7–1.8. It may be glimpsed in the dawn sky, especially later in the month. *Jupiter* is at superior conjunction on August 26. *Saturn* is at magnitude 0.4–0.5 throughout the month, and is still in *Libra*. It resumes direct motion on August 4 and is visible, early in the night in the southwest, setting about 23:30 (UT) early in the month.

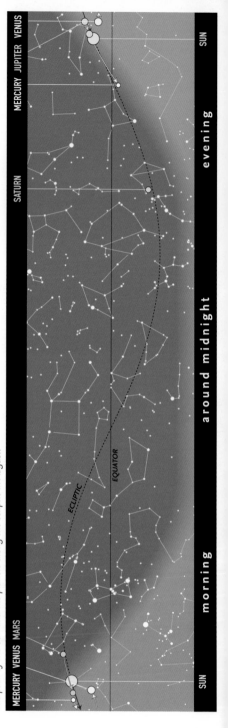

The path of the Sun and the planets along the ecliptic in August.

After midnight 2:00 (BST)

August 9 • The Moon and Aldebaran close together.

Morning 5:15 (BST)

August 13 • The crescent Moon and Mars, just before sunrise.

Evening 21:00 (BST)

August 22, 23 • The Moon passes Saturn and Antares

Morning 5:30 (BST)

August 29 • Venus and Mars in the eastern sky just

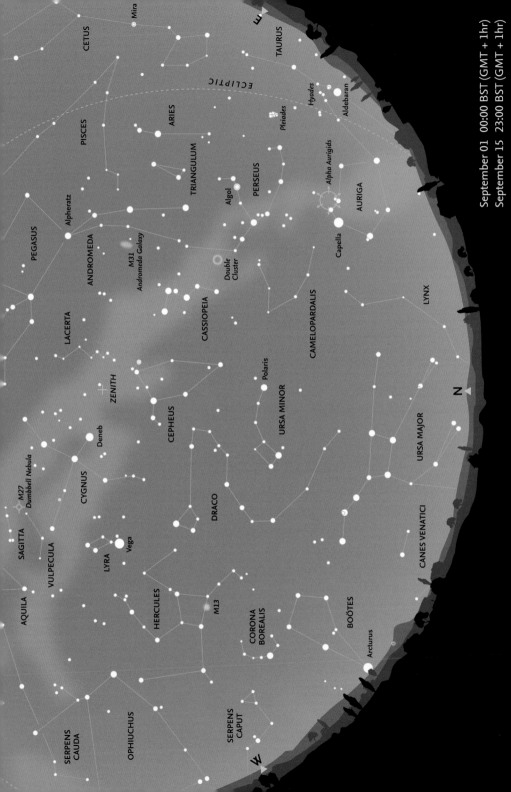

September 01   00:00 BST (GMT + 1hr)
September 15   23:00 BST (GMT + 1hr)

September – Looking North

Ursa Major is now low in the north and to the northwest *Arcturus* and much of *Boötes* sink below the horizon later in the night and later in the month. In the northeast *Auriga* is beginning to climb higher in the sky. Later in the month, *Taurus*, with orange *Aldebaran* (α Tauri), and even *Gemini*, with *Castor* and *Pollux*, become visible in the east and northeast. Due east, *Andromeda* is now clearly visible, with the small constellations of *Triangulum* and *Aries* (the latter a zodiacal constellation) directly below it. Practically the whole of the Milky Way is visible, arching across the sky, both in the north and in the south. It is not particularly clear in Auriga, or even *Perseus*, but in *Cassiopeia* and on towards *Cygnus* the clouds of stars become easier to see. *Cepheus* is 'upside-down' near the zenith, and the head of *Draco* and *Hercules* beyond it are well placed for observation.

Meteors

After the major Perseid shower in August, there is very little shower activity in September. One minor, but very extended, shower, known as the *Alpha Aurigids*, actually has two peaks of activity. The first was on August 28 but the primary peak occurs on September 15. At either of the maxima, however, the hourly rate hardly reaches ten meteors per hour, although the meteors are bright and relatively easy to photograph. Activity from this shower also extends into October. The *Southern Taurid* shower begins this month and, although rates are low, often produces very bright fireballs. As a slight compensation for the lack of activity, however, in September the number of sporadic meteors reaches its highest rate at any time during the year.

Total lunar eclipse of September 28

There is a total lunar eclipse on September 28, visible from western Europe, West Africa, eastern North America and the whole of South America. The eclipse begins at 00:12 UT, when the Moon first contacts the outer, faint shadow of the Earth (the penumbra). Only when the Moon contacts the inner, dark shadow (the umbra) at 01:07 UT will the eclipse become obvious. The whole Moon will be dark from 02:11 to 03:23 UT, with mid-eclipse at 02:48. The Moon will emerge completely from the umbra at 04:27 and from the penumbra at 05:22 UT. At mid-eclipse the Moon is fairly low in the southwest: at an altitude of about 27° for London and 26° for Edinburgh.

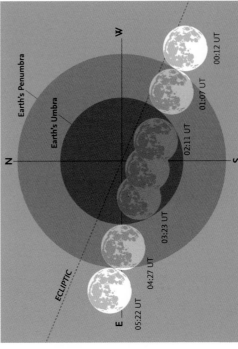

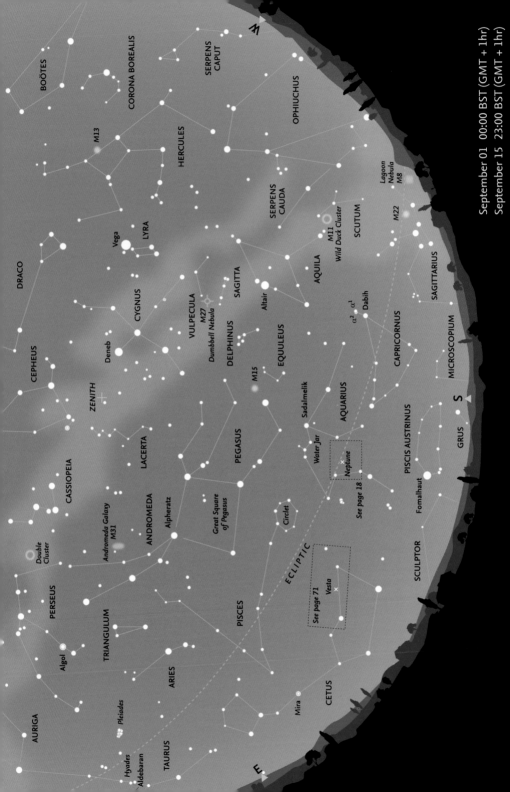

September 01 00:00 BST (GMT + 1hr)
September 15 23:00 BST (GMT + 1hr)

September – Looking South

The **Summer Triangle** is now high in the southwest, with the Great Square of **Pegasus** high in the south-east. Below Pegasus are the two zodiacal constellations of **Capricornus** and **Aquarius**. In what is otherwise an unremarkable constellation, α Capricorni is actually a visual binary, with the two stars (*Prima Giedi*, α¹ Cap and *Secunda Giedi*, α² Cap) readily seen with the naked eye. *Dabih* (β Capricorni), just to the south, is also a double star, and the components are relatively easy to separate with binoculars. In Aquarius, just to the east of *Sadalmelik* (α Aquarii) there is a small asterism consisting of four stars, resembling a tiny version of Cancer, known as the 'Water Jar'. Below Aquarius is a sparsely populated area of the sky with just one bright star in the constellation of **Piscis Austrinus**. In classical illustrations, water is shown flowing from the 'Water Jar' towards bright **Fomalhaut** (α Piscis Austrini).

Another zodiacal constellation, **Pisces**, is now clearly visible to the east of Aquarius. Although faint, there is a distinctive asterism of stars, known as the 'Circlet', south of the Great Square and another line of faint stars to the east of Pegasus. Still farther down towards the horizon is the constellation of **Cetus**, with the famous variable star **Mira** (ο Ceti) at its centre. When Mira is at maximum brightness (around mag. 3.5) it is clearly visible to the naked eye, but it disappears as it fades towards minimum (about mag. 9.5 or less).

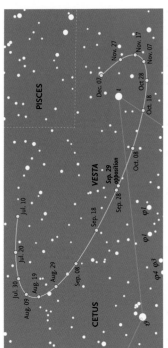

Minor planet Vesta comes to opposition on September 9 and is bright enough (mag. 6.2) to be detected with a pair of binoculars.

The Moon's phases for September

01	02	03	04	05 09:54	06
				Last Quarter	
07	08	09	10	11	12
13 06:41	14	15	16		
17	18	19	20	21 08:59	22
				First Quarter	
23	24	25	26	27	28 02:51
29	30				
New Moon	Full Moon				

September – Moon and Planets

The Moon

On September 5, the Last Quarter Moon is near **Aldebaran** (α Tauri) as it rises in the eastern sky. (Occultation in twilight at 04:50 UT for London and 04:51 UT for Edinburgh.) On September 10, the waning crescent Moon passes north of **Venus** and (in daylight) south of **Mars**. The following day it is south of **Regulus** (α Leonis) in the bright dawn sky. At New Moon on September 13, there is a partial solar eclipse, visible from South Africa, the southern Indian Ocean and Antarctica. A day later, the Moon is at apogee, the most distant of the year, at 406,464 km. On September 19 it passes north of **Saturn** in **Libra** and later that night north of **Antares** (α Scorpii). On September 28 there is the total lunar eclipse (described earlier) and the Moon comes to the closest perigee of the year at 356,876.769 km.

The Planets

Mercury reaches greatest eastern elongation on September 4 but is too low on the western horizon to be readily visible. **Venus**, visible in the morning sky, brightens slightly throughout the month from magnitude -4.5 to -4.7 as it moves from Cancer into Leo. **Mars** is at magnitude 1.8 throughout September. Initially it is in **Cancer** but moves into Leo on September 6. **Jupiter** (in **Leo**) is visible low in the east towards the end of the month at magnitude -1.7. **Saturn** is in **Libra** (magnitude 0.5–0.6), visible early in the month (and night) before it sets in the southwest. **Uranus** (magnitude 5.7) remains in **Pisces** and is visible throughout the night. On September 1, **Neptune** is at opposition in **Aquarius**. Although faint (magnitude 7.8) it should be visible in most binoculars (see chart on page 18).

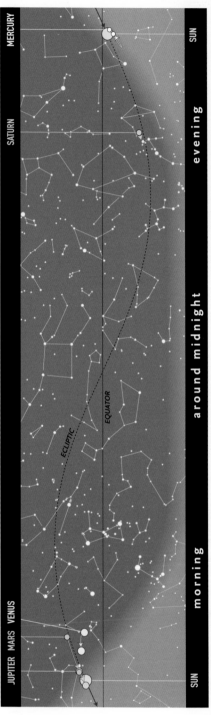

The path of the Sun and the planets along the ecliptic in September.

Calendar for September

Occultation of Aldebaran

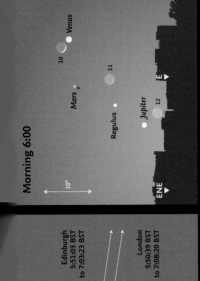

Edinburgh
5:51:03 BST
to 7:03:23 BST

London
5:50:39 BST
to 7:08:20 BST

N

S

September 5 • Twilight occultation of Aldebaran, as seen from London (bottom) and Edinburgh (top).

Evening 20:00 (BST)

10°

Moon

Antares

Saturn

SSW

SW

September 19 • The Moon with Saturn and Antares.

Morning 6:00

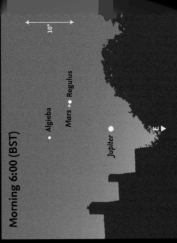

10°

Venus

10

Mars

11

Regulus

Jupiter

12

ENE

E

September 10–12 • The Moon passes Venus, Mars, Regulus and Jupiter.

Morning 6:00 (BST)

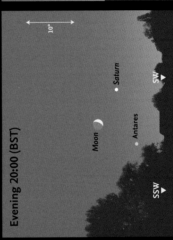

10°

Algieba

Mars • Regulus

Jupiter

E

September 25 • Mars and Regulus are close together.

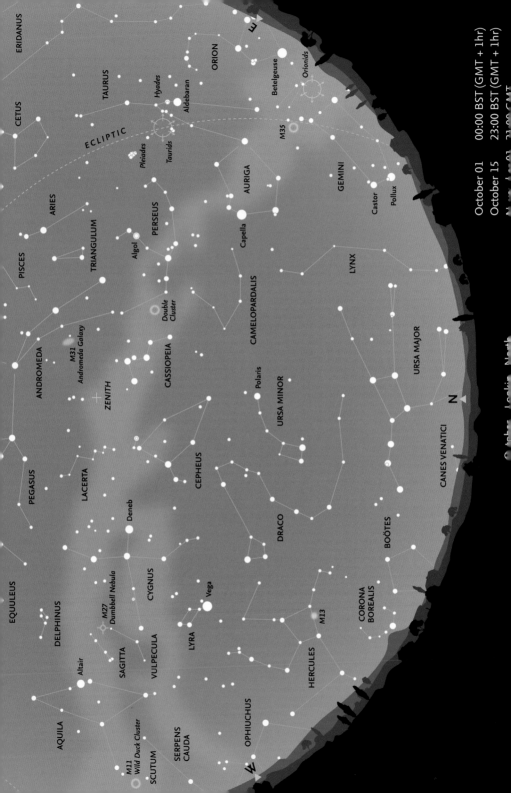

October 01   00:00 BST (GMT + 1hr)
October 15   23:00 BST (GMT + 1hr)

October – Looking North

The constellation of Cassiopeia is a familiar sight among the northern circumpolar constellations. It is always above the horizon, on the opposite side of Polaris (the North Star) to the equally well-known asterism of the seven stars of the Plough.

Ursa Major is grazing the horizon in the north, while high high overhead are the constellations of *Cepheus, Cassiopeia* and *Perseus*, with the Milky Way between Cepheus and Cassiopeia at the zenith. *Auriga* is now clearly visible in the east, as is *Taurus* with the *Pleiades, Hyades* and orange *Aldebaran*. Also in the east, *Orion* and *Gemini* are starting to rise clear of the horizon.

The constellations of *Boötes* and *Corona Borealis* are now essentially lost to view in the northwest, and *Hercules* is also descending towards the western horizon. The three stars of the Summer Triangle are still clearly visible, although *Aquila* and *Altair* are beginning to approach the horizon in the west. Towards the end of the month (October 25) Summer Time ends in Europe with Britain reverting to Greenwich Mean Time and Europe to Central European Time.

Meteors

The *Orionids* are the major, fairly reliable meteor shower active in October. Like the May *Eta Aquarid* shower, the Orionids are associated with Comet 1P/Halley. During this second pass through the stream of particles from the comet, slightly fewer meteors are seen than in May, but conditions are more favourable for northern observers. In both showers the meteors are very fast, and many leave persistent trains. Although the Orionid maximum is quoted as October 21–22, in fact there is a very broad maximum, lasting about a week from October 20 to 27, with hourly rates between 15 and 30. Occasionally rates are higher (50–70 per hour).

The faint shower of the *Southern Taurids* (often with bright fireballs) peaks on October 8–9. Towards the end of the month, another shower (the *Northern Taurids*) begins to show activity, which peaks early in November. The parent comet for both Taurid showers is Comet 2P/Encke.

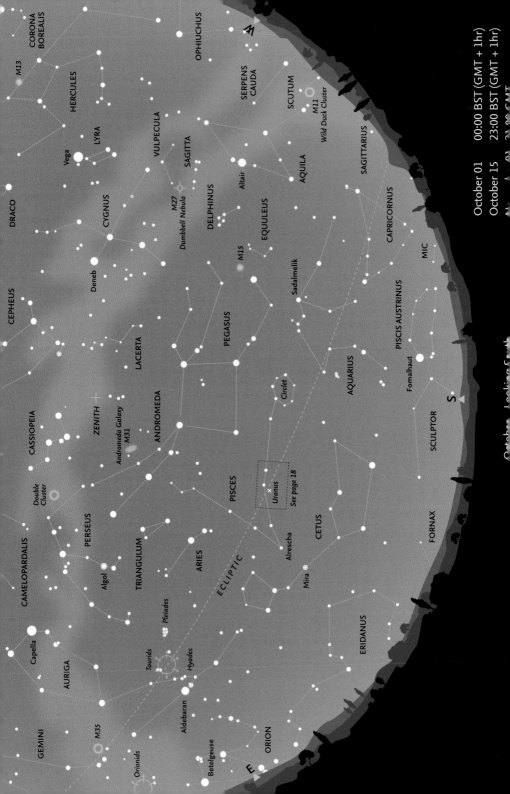

October 01   00:00 BST (GMT + 1hr)
October 15   23:00 BST (GMT + 1hr)

October Looking South

October – Looking South

The Great Square of *Pegasus* dominates the southern sky, framed by the two chains of stars that form the constellation of *Pisces*, together with *Alrescha* (α Piscium) at the point where the two lines of stars join. Also clearly visible is the constellation of *Cetus*, below Pegasus and Pisces. Although *Capricornus* is beginning to disappear, *Aquarius* to its east is well placed in the south, with solitary *Fomalhaut* and the constellation of *Piscis Austrinus* beneath it, close to the horizon.

The main band of the Milky Way and the Great Rift runs down from *Cygnus*, through *Vulpecula*, *Sagitta* and *Aquila* towards the western horizon. *Delphinus* and the tiny, unremarkable constellation of *Equuleus* lie between the band of the Milky Way and Pegasus. *Andromeda* is clearly visible high in the sky to the southeast, with the small constellation of *Triangulum* and the zodiacal constellation of *Aries* below it. *Perseus* is high in the east, and by now the *Pleiades* and *Taurus* are well clear of the horizon. Later in the night, and later in the month, *Orion* rises in the east, a sign that the autumn season has arrived and of the steady approach of winter.

The Moon's phases for October

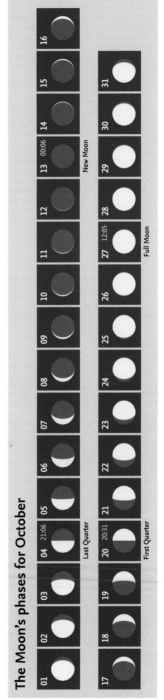

The constellation of Aquarius is one of the zodiacal constellations that is visible in late summer and early autumn. The four stars forming the 'Y'-shape of the 'Water Jar' may be seen to the east of Sadalmelik, the brightest star (top centre).

October – Moon and Planets

The Moon

On October 2, the waning gibbous Moon passes due north of *Aldebaran* (α Tauri) in daylight. On October 8 and 9, it is visible in the dawn sky close to *Venus*, *Mars*, *Jupiter* and *Regulus* (α Leonis). On October 11, it is close to *Mercury* as the planet moves towards greatest elongation (on October 16). On October 26, the Moon passes Uranus (in Pisces), but it is a day before Full Moon, so the planet will be lost in moonlight. On October 29, two days after Full Moon, it occults *Aldebaran* (α Tauri) at 23:07 UT. The event is visible over a large part of Europe, including Britain and Ireland.

The Planets

Throughout the month, three planets: *Venus* (magnitude -4.4–4.5), *Mars* (magnitude 1.8) and *Jupiter* (magnitude -1.6) are all clustered together in the dawn sky, near *Regulus* (α Leonis). *Mercury* joins them mid-month, reaching eastern elongation on October 16. *Venus* reaches western elongation on October 26, when it is very bright (magnitude -4.5). Early in the month it may be possible to glimpse *Saturn* low in the southwest just before it sets, but it is invisible for most of October. On October 12, *Uranus* reaches opposition in Pisces at magnitude 5.7, just a day before New Moon (see chart on page 18). It is visible throughout the night, and sets around 05:00 UT at the end of October.

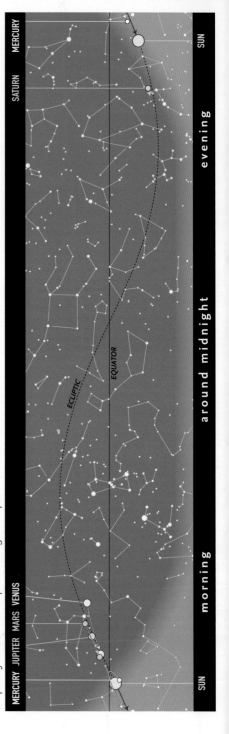

The path of the Sun and the planets along the ecliptic in October.

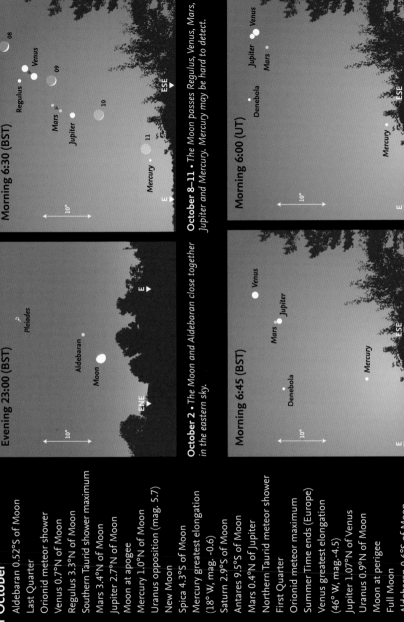

Calendar for October

02	13:14	Aldebaran 0.52°S of Moon
04	21:06	Last Quarter
Oct.04–Nov.14		Orionid meteor shower
08	20:32	Venus 0.7°N of Moon
08	22:17	Regulus 3.3°N of Moon
08–09		Southern Taurid shower maximum
09	16:51	Mars 3.4°N of Moon
09	23:34	Jupiter 2.7°N of Moon
11	13:18	Moon at apogee
11	12:00	Mercury 1.0°N of Moon
12	03:49	Uranus opposition (mag. 5.7)
13	00:06	New Moon
13	06:26	Spica 4.3°S of Moon
16	03:16	Mercury greatest elongation (18° W, mag. -0.6)
16	12:59	Saturn 2.9°S of Moon
17	01:55	Antares 9.5°S of Moon
17	13:00 *	Mars 0.4°N of Jupiter
Oct.19–Dec.10		Northern Taurid meteor shower
20	20:31	First Quarter
21–22		Orionid meteor maximum
25		Summer Time ends (Europe)
26	07:11	Venus greatest elongation (46° W, mag.-4.5)
26	08:00 *	Jupiter 1.07°N of Venus
26	10:24	Uranus 0.9°N of Moon
26	13:01	Moon at perigee
27	12:05	Full Moon
29	23:07	Aldebaran 0.6°S of Moon

* These objects are close together for an extended period around this time.

Evening 23:00 (BST)

October 2 • The Moon and Aldebaran close together in the eastern sky.

Morning 6:30 (BST)

October 8–11 • The Moon passes Regulus, Venus, Mars, Jupiter and Mercury. Mercury may be hard to detect.

Morning 6:45 (BST)

October 16 • Four planets in the morning sky. Jupiter and Mars are in conjunction at 13:00, during daylight.

Morning 6:00 (UT)

October 26 • The same four planets again. Venus and Jupiter are in conjunction at 8:00 (UT), after sunrise.

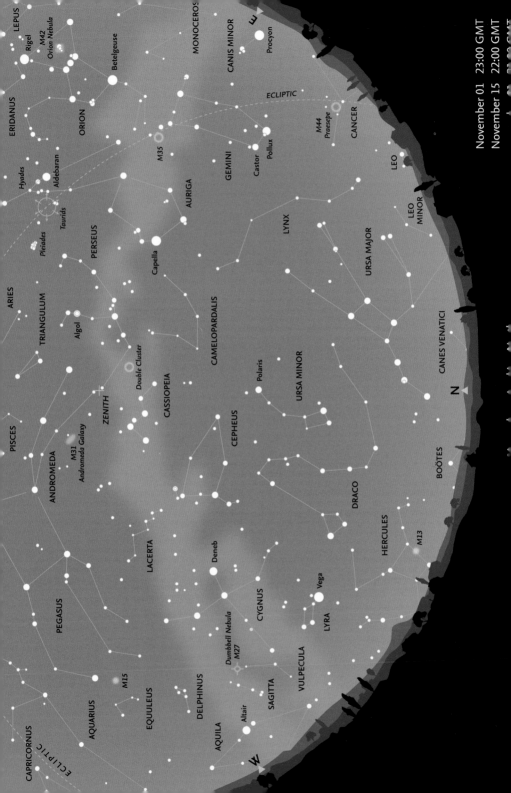

November 01 23:00 GMT
November 15 22:00 GMT

November – Looking North

Aquila has now disappeared below the horizon, but two of the stars of the Summer Triangle, **Vega** in **Lyra** and **Deneb** in **Cygnus**, are still visible in the west. The head of **Draco** is now low in the northwest and only a small portion of **Hercules** remains above the horizon. The Milky Way arches overhead, with the denser star clouds in the west and the less heavily populated region through **Auriga** and **Monoceros** in the east. High overhead, **Cassiopeia** is near the zenith and **Cepheus** has swung round to the northwest, while Auriga is now high in the northeast. **Gemini**, with **Castor** and **Pollux**, is well clear of the eastern horizon, and even **Procyon** (α Canis Minoris) is just climbing into view almost due east.

Meteors

The **Northern Taurid** shower, which began in mid-October, reaches maximum – although with only a low rate of about five meteors per hour – on November 12–13, but gradually trails off, ending around December 10. There is an apparent seven-year periodicity in fireball activity, and 2015 may be another peak year. Far more striking, however, are the **Leonids**, which have a moderate period of activity (November 5–30), with maximum on November 17–18. This shower is associated with Comet 55P/Tempel-Tuttle and has shown extraordinary activity on various occasions with many thousands of meteors per hour. High rates were seen in 1999, 2001 and 2002 (reaching about 3000 meteors per hour) but have fallen dramatically since then. The rate in 2015 is likely to be about 15 per hour. The meteors are the fastest shower meteors recorded (about 70 km/s) and often leave persistent trains. The shower is very rich in faint meteors. In 2015, maximum is just before First Quarter (November 19), so the waxing crescent should not create too much interference. Observing conditions are always best after midnight.

The constellation of Auriga, with Capella its brightest star (top right), and the distinctive triangle of the 'Kids' to its west (centre right) are prominent features in the northern sky. Capella itself is circumpolar for latitude 50° north and skims the horizon during the summer, but never actually disappears below it.

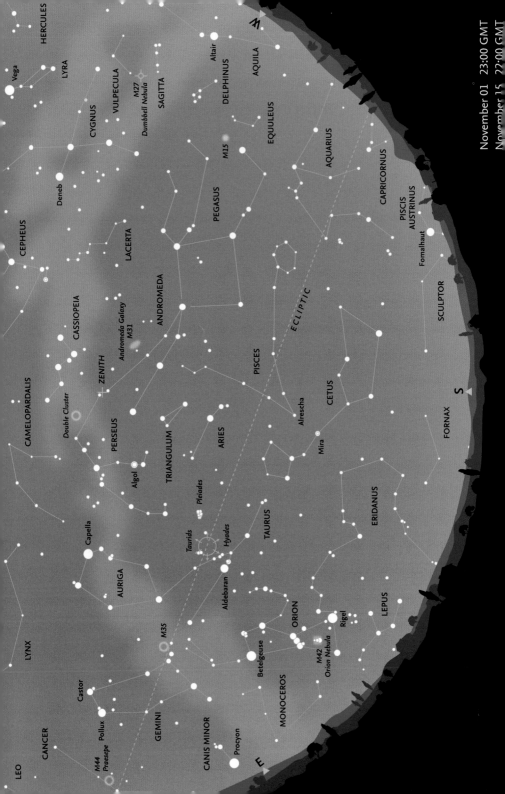

November – Looking South

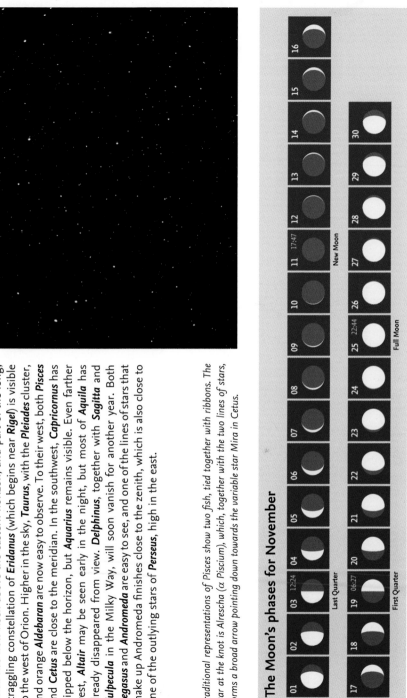

Orion has now risen above the eastern horizon, and part of the long, straggling constellation of **Eridanus** (which begins near **Rigel**) is visible to the west of Orion. Higher in the sky, **Taurus**, with the **Pleiades** cluster, and orange **Aldebaran** are now easy to observe. To their west, both **Pisces** and **Cetus** are close to the meridian. In the southwest, **Capricornus** has slipped below the horizon, but **Aquarius** remains visible. Even farther west, **Altair** may be seen early in the night, but most of **Aquila** has already disappeared from view. **Delphinus**, together with **Sagitta** and **Vulpecula** in the Milky Way, will soon vanish for another year. Both **Pegasus** and **Andromeda** are easy to see, and one of the lines of stars that make up Andromeda finishes close to the zenith, which is also close to one of the outlying stars of **Perseus**, high in the east.

Traditional representations of Pisces show two fish, tied together with ribbons. The star at the knot is Alrescha (α Piscium), which, together with the two lines of stars, forms a broad arrow pointing down towards the variable star Mira in Cetus.

The Moon's phases for November

| 01 | 02 | 03 12:24 | 04 | 05 | 06 | 07 | 08 | 09 | 10 | 11 17:47 | 12 | 13 | 14 | 15 | 16 |

Last Quarter

New Moon

| 17 | 18 | 19 06:27 | 20 | 21 | 22 | 23 | 24 | 25 22:44 | 26 | 27 | 28 | 29 | 30 |

First Quarter

Full Moon

November – Moon and Planets

The Moon

On November 1–2, the waning gibbous Moon passes close to the *Pleiades* and *Aldebaran* (α Tauri). A few days later, on November 5–7, the waning crescent passes, in sequence, south of *Regulus* (α Leonis), *Jupiter*, *Mars* and *Venus*. On November 9, two days before New, the Moon passes north of Spica (α Virginis) in daylight, but the two bodies may be glimpsed in the east, just before dawn. For most of the month, lunar appulses and conjunctions are lost in daylight. On November 22, the waxing gibbous Moon passes about 1° south of *Uranus* (magnitude 5.7) in *Pisces*. On November 25–26, the Moon passes well south of the *Pleiades* and then just north of *Aldebaran* (α Tauri).

The Planets

Venus, *Mars* and *Jupiter* remain clustered together in the eastern sky early in the month. *Venus* moves back towards the Sun (following western elongation on October 26), but *Mars* and *Jupiter* rise higher as the month progresses. *Venus* may be seen close to *Spica* (α Virginis) in the pre-dawn sky on November 28. *Mercury* is very low in the dawn sky early in the month and essentially impossible to detect. Saturn sets shortly after the Sun and is invisible. *Uranus* (in *Pisces* – see chart on page 18) remains visible at magnitude 5.7 throughout the night.

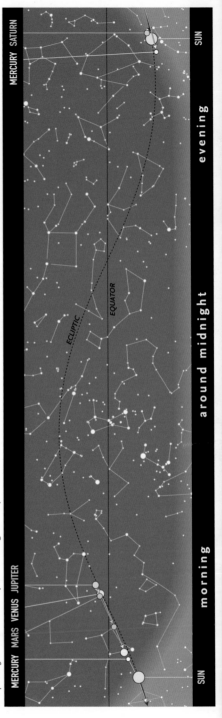

The path of the Sun and the planets along the ecliptic in November.

Calendar for November

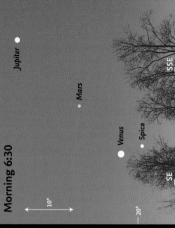

Morning 6:00

November 5–8 • The Moon passes Regulus, Jupiter, Mars and Venus.

November 9 • The thin crescent Moon and Spica, just before dawn.

November 26 • The Full Moon with Aldebaran near the western horizon.

November 28 • Venus with Spica in the morning sky. Mars and Jupiter are a little higher.

01		Daylight Saving Time ends (North America)
03	12:24	Last Quarter
03	16:00 *	Mars 0.7°N of Venus
05	04:44	Regulus 3.2°N of Moon
05–30		Leonid meteor shower
06	15:53	Jupiter 2.3°N of Moon
07	09:57	Mars 1.8°N of Moon
07	13:54	Venus 1.2°N of Moon
07	21:50	Moon at apogee
09	12:52	Spica 4.3°S of Moon
11	07:39	Mercury 3.2°S of Moon
11	17:47	New Moon
12–13		Northern Taurid shower maximum
13	00:45	Saturn 3.0°S of Moon
13	07:43	Antares 9.4°S of Moon
17–18		Leonid meteor maximum
19	06:27	First Quarter
22	18:41	Uranus 0.9°N of Moon
23	20:07	Moon at perigee
25	22:44	Full Moon
26	09:55	Aldebaran 0.7°S of Moon
28	16:00 *	Spica 4.5°S of Venus

* These objects are close together for an extended period around this time.

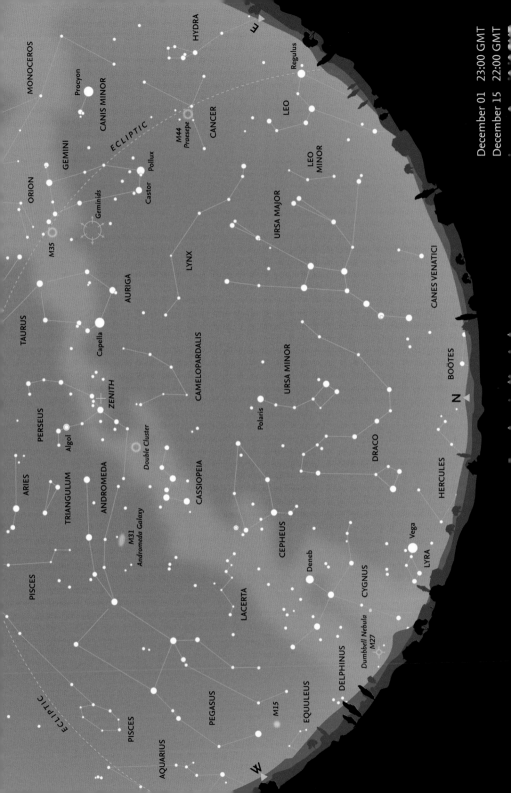

MONOCEROS

HYDRA

E

CANIS MINOR

Procyon

GEMINI

Regulus

ORION

ECLIPTIC

M44
Praesepe

CANCER

LEO

Geminids

Pollux

Castor

LEO
MINOR

M35

URSA MAJOR

TAURUS

AURIGA

LYNX

Capella

CAMELOPARDALIS

CANES VENATICI

ZENITH

PERSEUS

Algol

URSA MINOR

BOÖTES

ARIES

Double Cluster

Polaris

N

ANDROMEDA

CASSIOPEIA

DRACO

TRIANGULUM

HERCULES

M31
Andromeda Galaxy

CEPHEUS

PISCES

Vega

LYRA

Deneb

LACERTA

CYGNUS

PEGASUS

PISCES

Dumbbell Nebula
M27

AQUARIUS

M15

EQUULEUS

DELPHINUS

ECLIPTIC

W

December 01   23:00 GMT
December 15   22:00 GMT

December – Looking North

The constellation of Orion dominates the sky during this period of the year, and is a useful starting point for recognizing other constellations in the southern sky. This photograph was taken with a 25-second exposure on an ordinary camera. Orion can be found in the southern part of the sky (see next page).

Ursa Major has now swung round and is starting to 'climb' in the east. The fainter stars in the southern part of the constellation are now fully in view. The other bear, **Ursa Minor**, 'hangs' below **Polaris** in the north. Directly above it is the faint constellation of **Camelopardalis**, with the other inconspicuous circumpolar constellation, **Lynx**, to its east. **Vega** (α Lyrae) is skimming the horizon in the northwest, but **Deneb** (α Cygni) and most of **Cygnus** remain visible farther west. In the east, **Regulus** (α Leonis) and the constellation of **Leo** are beginning to rise above the horizon. **Cancer** stands high in the east, with **Gemini** even higher in the sky. **Perseus** is at the zenith, with **Auriga** and **Capella** between it and Gemini. Because it is so high in the sky, now is a good time to examine the star clouds of the fainter portion of the Milky Way, between **Cassiopeia** in the west to Gemini and **Orion** in the east.

Meteors

There is one significant meteor shower in December (the last major shower of the year). This is the **Geminid** shower, which is visible over the period December 4–16 and comes to maximum on December 13–14. It is one of the most active showers of the year, and in some years is the most active, with a peak rate of around 100 meteors per hour. It is the one major shower that shows good activity before midnight. The meteors have been found to have a much higher density than other meteors (which are derived from cometary material). It was eventually established that the Geminids and the asteroid Phaeton had similar orbits. So the Geminids are assumed to consist of denser, rocky material. They are slower than most other meteors and often appear to last longer. The brightest often break up into numerous luminous fragments that follow similar paths across the sky. There is a second shower: the **Ursids**, active December 17–23, peaking on December 21–22, with rate at maximum of 5–10, occasionally rising to 25 per hour. The parent body is Comet 8P/Tuttle.

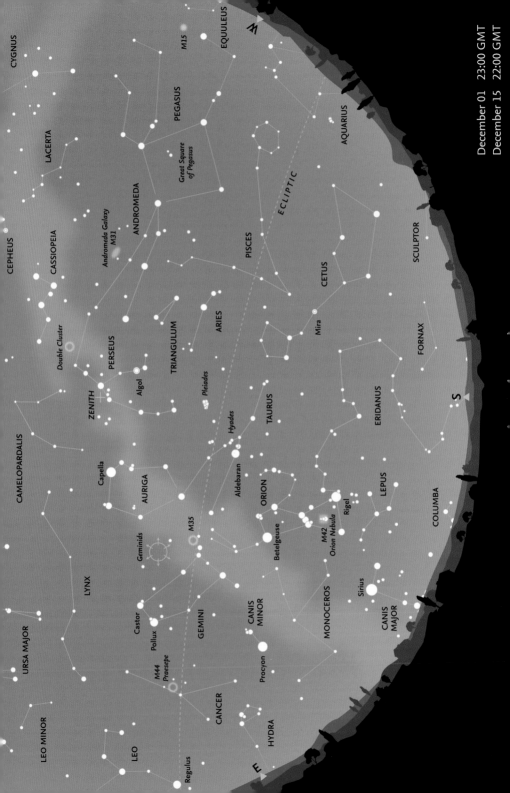

December 01   23:00 GMT
December 15   22:00 GMT

December – Looking South

The fine open cluster of the *Pleiades* is due south around 22:00, with the *Hyades* cluster, *Aldebaran* and the rest of *Taurus* clearly visible to the east. *Auriga* (with *Capella*) and *Gemini* (with *Castor* and *Pollux*) are both well-placed for observation. *Orion* has made a welcome return to the winter sky, and both *Canis Minor* (with *Procyon*) and *Canis Major* (with *Sirius*, the brightest star in the sky) are now well above the horizon. The small, poorly known constellation of *Lepus* lies to the south of Orion. In the east, *Aquarius* has now disappeared, and *Cetus* is becoming lower, but *Pisces* is still easily seen, as are the constellations of *Aries*, *Triangulum* and *Andromeda* above it. The Great Square of *Pegasus* is starting to plunge down towards the western horizon, and because of its orientation on the sky appears more like a large diamond, standing on one point, than a square.

The constellation of Taurus contains two contrasting open clusters: the compact Pleiades, with its striking blue-white stars, and the more scattered, but much closer,'V'-shaped Hyades. Orange Aldebaran (α Tauri) is not related to the Hyades, but lies between it and the Earth.

The Moon's phases for December

01
02
03 07:40 Last Quarter
04
05
06
07
08
09
10
11 10:29 New Moon
12
13
14
15
16

17
18 15:14 First Quarter
19
20
21
22
23
24
25 11:11 Full Moon
26
27
28
29
30
31

December – Moon and Planets

The Moon

Early in the month, the waning Moon passes close to **Regulus** (α Leonis) and **Jupiter**. A few days after Full Moon on December 25, it passes the star and planet for a second time. On December 5, the waning crescent will pass close to **Mars**. There will be an occultation, visible only from northeastern Africa. **Aldebaran** (α Tauri) is occulted by the Moon on December 23 (two days before Full). This event is visible from the whole of Europe and western North Africa. The time given (19:32) is the exact time of conjunction, but the occultation will begin about 20 minutes earlier for the UK and Ireland, with the star disappearing at the dark (eastern) limb of the Moon.

The Planets

Mercury is close to the Sun, but reaches eastern elongation on December 29, when it may be visible low in the southwest shortly after sunset. Early in the month, **Venus** is in **Virgo** in the morning sky, not far from **Spica**. It moves into **Libra** by the end of the month. **Mars** is in **Virgo**, in the southeastern sky, lying almost due south by dawn at the end of the month. It brightens slightly from magnitude 1.5 to 1.3 over the month. **Jupiter** is in **Leo** throughout the month, moving towards the border of **Virgo** by the end of the year. Initially at magnitude -2.0 it brightens slightly to -2.2 at the end of December. **Saturn** is initially far too close to the Sun to be visible, but it may be glimpsed in the southeast shortly before sunrise at the end of the month. **Uranus** remains in **Pisces**, at magnitude 5.8.

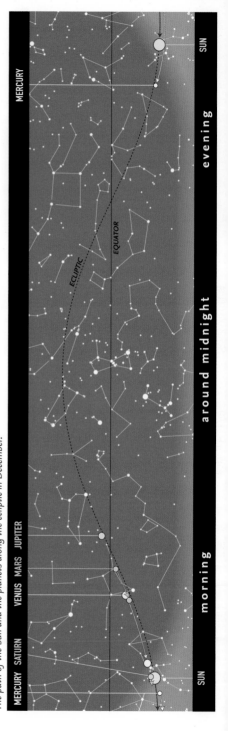

The path of the Sun and the planets along the ecliptic in December.

Calendar for December

02	12:19	Regulus 3.0°N of Moon
03	07:40	Last Quarter
04–16		Geminid meteor shower
04	06:26	Jupiter 1.8°N of Moon
05	14:56	Moon at apogee
06	02:42	Mars 0.1°N of Moon (occultation NE Africa)
06	20:08	Spica 4.5°S of Moon
07	16:56	Venus 0.7°S of Moon
0	14:07	Saturn 3.1°S of Moon
0	14:44	Moon 9.4°N of Antares
1	10:29	New Moon
2	14:09	Mercury 7.2°S of Moon
3	08:00 *	Saturn 6.27°N of Antares
3–14		Geminid shower maximum
7–23		Ursid meteor shower
8	15:14	First Quarter
0	00:50	Uranus 1.2°N of Moon
1	09:00	Moon at perigee
1	12:00 *	Mars 3.8°N of Spica
21–22		Ursid shower maximum
2	04:48	Winter solstice
3	19:32	Aldebaran 0.7°S of Moon (occultation from UK)
25	11:11	Full Moon
29	03:12	Mercury greatest elongation (20° E, mag. –0.6)
9	20:55	Regulus 2.7°N of Moon
1	18:00	Jupiter 1.5°N of Moon

* These objects are close together for an extended

Morning 7:00

December 4–8 • *The Moon passes Jupiter, Mars, Spica and Venus.*

Morning 7:00

December 21 • *Mars and Spica are close together.*

Morning 7:00

29 December 2015–1 January 2016 • *The Moon passes*

Glossary and Tables

aphelion	The point on an orbit that is farthest from the Sun.
apogee	The point on its orbit at which the Moon is farthest from the Earth.
appulse	The apparently close approach of two celestial objects; two planets, or a planet and star.
astronomical unit	(AU) The mean distance of the Earth from the Sun, 149,597,870 km.
celestial equator	The great circle on the celestial sphere that is in the same plane as the Earth's equator.
celestial sphere	The apparent sphere surrounding the Earth on which all celestial bodies (stars, planets, etc.) seem to be located.
conjunction	The point in time when two celestial objects have the same celestial longitude. In the case of the Sun and a planet, superior conjunction occurs when the planet lies on the far side of the Sun (as seen from Earth). For Mercury and Venus, inferior conjuction occurs when they pass between the Sun and the Earth.
direct motion	Motion from west to east on the sky.
ecliptic	The apparent path of the Sun across the sky throughout the year. Also: the plane of the Earth's orbit in space.
elongation	The point at which an inferior planet has the greatest angular distance from the Sun, as seen from Earth.
equinox	The two points during the year when night and day have equal duration. Also: the points on the sky at which the ecliptic intersects the celestial equator. The vernal (spring) equinox is of particular importance in astronomy.
gibbous	The stage in the sequence of phases at which the illumination of a body lies between half and full. In the case of the Moon, the term is applied to phases between First Quarter and Full, and between Full and Last Quarter.
inferior planet	Either of the planets Mercury or Venus, which have orbits inside that of the Earth.
magnitude	The brightness of a star, planet or other celestial body. It is a logarithmic scale, where larger numbers indicate fainter brightness. A difference of 5 in magnitude indicates a difference of 100 in actual brightness, thus a first-magnitude star is 100 times as bright as one of sixth magnitude.
meridian	The great circle passing through the North and South Poles of a body and the observer's position; or the corresponding great circle on the celestial sphere that passes through the North and South Celestial Poles and also through the observer's zenith.
nadir	The point on the celestial sphere directly beneath the observer's feet, opposite the zenith.
occultation	The disappearance of one celestial body behind another, such as when stars or planets are hidden behind the Moon.
opposition	The point on a superior planet's orbit at which it is directly opposite the Sun in the sky.
perigee	The point on its orbit at which the Moon is closest to the Earth.
perihelion	The point on an orbit that is closest to the Sun.
retrograde motion	Motion from east to west on the sky.
superior planet	A planet that has an orbit outside that of the Earth.
vernal equinox	The point at which the Sun, in its apparent motion along the ecliptic, crosses the celestial equator from south to north. Also known as the First Point of Aries.
zenith	The point directly above the observer's head.
zodiac	A band, 8° on either side of the ecliptic, within which the Moon and planets appear to move. It consists of twelve equal areas, originally named after the constellation that once lay within it.

The Greek Alphabet

α	Alpha	ε	Epsilon	ι	Iota	ν	Nu	ρ	Rho	φ (φ)	Phi
β	Beta	ζ	Zeta	κ	Kappa	ξ	Xi	σ (ς)	Sigma	χ	Chi
γ	Gamma	η	Eta	λ	Lambda	o	Omicron	τ	Tau	ψ	Psi
δ	Delta	θ (ϑ)	Theta	μ	Mu	π	Pi	υ	Upsilon	ω	Omega

The Constellations

There are 88 constellations covering the whole of the celestial sphere, but 24 of these in the southern hemisphere can never be seen (even in part) from the latitude of Britain and Ireland, so are omitted from this table. The names themselves are expressed in Latin, and the names of stars are frequently given by Greek letters followed by the genitive of the constellation name. The genitives and English names of the various constellations are included.

Name	Genitive	Abbr.	English name
Andromeda	Andromeda	And	Andromeda
Antlia	Antliae	Ant	Air Pump
Aquarius	Aquarii	Aqr	Water Bearer
Aquila	Aquilae	Aql	Eagle
Aries	Arietis	Ari	Ram
Auriga	Aurigae	Aur	Charioteer
Boötes	Boötis	Boo	Herdsman
Camelopardalis	Camelopardalis	Cam	Giraffe
Cancer	Cancri	Cnc	Crab
Canes Venatici	Canum Venaticorum	CVn	Hunting Dogs
Canis Major	Canis Majoris	CMa	Big Dog
Canis Minor	Canis Minoris	CMi	Little Dog
Capricornus	Capricorni	Cap	Sea Goat
Cassiopeia	Cassiopeiae	Cas	Cassiopeia
Centaurus	Centauri	Cen	Centaur
Cepheus	Cephei	Cep	Cepheus
Cetus	Ceti	Cet	Whale
Columba	Columbae	Col	Dove
Coma Berenices	Coma Berenicis	Com	Berenice's Hair
Corona Australis	Coronae Australis	CrA	Southern Crown
Corona Borealis	Coronae Borealis	CrB	Northern Crown
Corvus	Corvi	Crv	Crow
Crater	Crateris	Crt	Cup
Cygnus	Cygni	Cyg	Swan
Delphinus	Delphini	Del	Dolphin
Draco	Draconis	Dra	Dragon
Equuleus	Equulei	Equ	Little Horse
Eridanus	Eridani	Eri	River Eridanus
Fornax	Fornacis	For	Furnace
Gemini	Geminorum	Gem	Twins
Hercules	Herculis	Her	Hercules
Hydra	Hydrae	Hya	Water Snake

Name	Genitive	Abbr.	English name
Lacerta	Lacertae	Lac	Lizard
Leo	Leonis	Leo	Lion
Leo Minor	Leonis Minoris	LMi	Little Lion
Lepus	Leporis	Lep	Hare
Libra	Librae	Lib	Scales
Lupus	Lupi	Lup	Wolf
Lynx	Lyncis	Lyn	Lynx
Lyra	Lyrae	Lyr	Lyre
Microscopium	Microscopii	Mic	Microscope
Monoceros	Monocerotis	Mon	Unicorn
Ophiuchus	Ophiuchi	Oph	Serpent Bearer
Orion	Orionis	Ori	Orion
Pegasus	Pegasi	Peg	Pegasus
Perseus	Persei	Per	Perseus
Pisces	Piscium	Psc	Fishes
Piscis Austrinus	Piscis Austrini	PsA	Southern Fish
Puppis	Puppis	Pup	Stern
Pyxis	Pyxidis	Pyx	Compass
Sagitta	Sagittae	Sge	Arrow
Sagittarius	Sagittarii	Sgr	Archer
Scorpius	Scorpii	Sco	Scorpion
Sculptor	Sculptoris	Scl	Sculptor
Scutum	Scuti	Sct	Shield
Serpens	Serpentis	Ser	Serpent
Sextans	Sextantis	Sex	Sextant
Taurus	Tauri	Tau	Bull
Triangulum	Trianguli	Tri	Triangle
Ursa Major	Ursae Majoris	UMa	Great Bear
Ursa Minor	Ursae Minoris	UMi	Lesser Bear
Vela	Velorum	Vel	Sail
Virgo	Virginis	Vir	Virgin
Vulpecula	Vulpeculae	Vul	Fox

Some common asterisms

Belt of Orion	δ, ε, and ζ Orionis
Big Dipper	α, β, γ, δ, ε, ζ, and η Ursae Majoris
Circlet	γ, θ, ι, λ, and κ Piscium
Guards (or Guardians)	β and γ Ursae Minoris
Head of Cetus	α, γ, ξ², μ, and λ Ceti
Head of Draco	β, γ, ξ, and ν Draconis
Head of Hydra	δ, ε, ζ, η, ρ, and σ Hydrae
Keystone	ε, ζ, η, and π Herculis
Kids	ε, ζ, and η Aurigae
Little Dipper	β, γ, η, ζ, ε, δ, and α Ursae Minoris
Lozenge	= Head of Draco
Milk Dipper	ζ, γ, σ, φ, and λ Sagittarii
Plough	α, β, γ, δ, ε, ζ, and η Ursae Majoris
Pointers	α and β Ursae Majoris
Sickle	α, η, γ, ζ, μ, and ε Leonis
Square of Pegasus	α, β, and γ Pegasi with α Andromedae
Sword of Orion	θ and ι Orionis
Teapot	γ, ε, δ, λ, φ, σ, τ, and ζ Sagittarii
Wain (or Charles' Wain)	= Plough
Water Jar	γ, η, κ, and ζ Aquarii
Y of Aquarius	= Water Jar

Acknowledgements

Martin Koitmäe, Estonia: p.51 (NLC image)

Storm Dunlop, Chichester: p.87 (Orion)

Steve Edberg, La Cañada, California: all other constellation photographs

Eclipse Predictions by Fred Espenak, NASA/GSFC: p.33 (Eclipse diagram)

Editorial support was provided by Tom Kerss, Planetarium Astronomer at the Royal Observatory Greenwich, part of Royal Museums Greenwich.

Further Information

Books

Bone, Neil (1993), *Observer's Handbook: Meteors*, George Philip, London & Sky Publ. Corp., Cambridge, Mass.

Cook, J., ed. (1999), *The Hatfield Photographic Lunar Atlas*, Springer-Verlag, New York

Dunlop, Storm (1999), *Wild Guide to the Night Sky*, HarperCollins, London

Dunlop, Storm (2012), *Practical Astronomy*, 3rd edn, Philip's, London

Dunlop, Storm, Rükl, Antonin & Tirion, Wil (2005), *Collins Atlas of the Night Sky*, HarperCollins, London

Ridpath, Ian, ed. (1998), *Norton's Star Atlas*, 19th edn., Longman, London

Ridpath, Ian, ed. (2003), *Oxford Dictionary of Astronomy*, 2nd edn, Oxford University Press, Oxford

Ridpath, Ian & Tirion, Wil (2004), *Collins Gem - Stars*, HarperCollins, London

Ridpath, Ian & Tirion, Wil (2011), *Collins Pocket Guide Stars and Planets*, 4th edn, HarperCollins, London

Ridpath, Ian & Tirion, Wil (2012), *Monthly Sky Guide*, 9th edn, Cambridge University Press

Rükl, Antonín (1990), *Hamlyn Atlas of the Moon*, Hamlyn, London & Astro Media Inc., Milwaukee

Rükl, Antonín (2004), *Atlas of the Moon*, Sky Publishing Corp., Cambridge, Mass.

Scagell, Robin (2000), *Philip's Stargazing with a Telescope*, George Philip, London

Tirion, Wil (2011), *Cambridge Star Atlas*, 4th edn, Cambridge University Press, Cambridge

Tirion, Wil & Sinnott, Roger (1999), *Sky Atlas 2000.0*, 2nd edn, Sky Publishing Corp., Cambridge, Mass.
& Cambridge University Press, Cambridge

Journals

Astronomy, Astro Media Corp., 21027 Crossroads Circle, P.O. Box 1612, Waukesha,
WI 53187-1612 USA. http://www.astronomy.com

Astronomy Now, Pole Star Publications, PO Box 175, Tonbridge, Kent TN10 4QX UK.
http://www.astronomynow.com

Sky & Telescope, Sky Publishing Corp., Cambridge, MA 02138-1200, USA.
http://www.skyandtelescope.com/

Societies

British Astronomical Association, Burlington House, Piccadilly, London W1J 0DU.
http://www.britastro.org/
The principal British organization for amateur astronomers (with some professional members), particularly for those interested in carrying out observational programmes. Its membership is, however, worldwide. It publishes fully refereed, scientific papers and other material in its well-regarded Journal.

Federation of Astronomical Societies, Secretary: Ken Sheldon, Whitehaven, Maytree Road, Lower Moor, Pershore, Worcs. WR10 2NY. http://www.fedastro.org.uk/fas/
An organization that is able to provide contact information for local astronomical societies in the United Kingdom.

Royal Astronomical Society, Burlington House, Piccadilly, London W1J 0BQ. http://www.ras.org.uk/
The premier astronomical society, with membership primarily drawn from professionals and experienced amateurs. It has an exceptional library and is a designated centre for the retention of certain classes of astronomical data. Its publications are the standard medium for dissemination of astronomical research.

Society for Popular Astronomy, 36 Fairway, Keyworth, Nottingham NG12 5DU.
http://www.popastro.com/
A society for astronomical beginners of all ages, which concentrates on increasing members' understanding and enjoyment, but which does have some observational programmes. Its journal is entitled *Popular Astronomy*.

Software

Planetary, Stellar and Lunar Visibility, (Planetary and eclipse freeware): Alcyone Software, Germany.
http://www.alcyone.de
Redshift, Redshift-Live. http://www.redshift-live.com/en/
Starry Night & Starry Night Pro, Sienna Software Inc., Toronto, Canada. http://www.starrynight.com

Internet sources

There are numerous sites with information about all aspects of astronomy, and all of those have numerous links. Although many amateur sites are excellent, treat any statements and data with caution. The sites listed below offer accurate information. Please note that the URLs may change. If so, use a good search engine, such as Google, to locate the information source.

Information

Auroral information Michigan Tech: http://www.geo.mtu.edu/weather/aurora/
Comets JPL Solar System Dynamics: http://ssd.jpl.nasa.gov/
American Meteor Society http://amsmeteors.org/
Deep-sky objects Saguaro Astronomy Club Database: http://www.virtualcolony.com/sac/
Eclipses: NASA Eclipse Page: http://eclipse.gsfc.nasa.gov/eclipse.html
Moon (inc. Atlas) Inconstant Moon: http://www.inconstantmoon.com/
Planets Planetary Fact Sheets: http://nssdc.gsfc.nasa.gov/planetary/planetfact.html
Satellites (inc. International Space Station)
Heavens Above: http://www.heavens-above.com/
Visual Satellite Observer's: http://www.satobs.org/
Star Chart National Geographic Chart
http://www.nationalgeographic.com/features/97/stars/chart/index.html
What's Visible Skyhound: http://www.skyhound.com/sh/skyhound.html
Skyview Cafe: http://www.skyviewcafe.com
Stargazer: http://www.outerbody.com/stargazer/ (choose 'File, New View', click & drag to change)

Institutes and Organizations

European Space Agency: http://www.esa.int/
International Dark-Sky Association: http://www.darksky.org/
Jet Propulsion Laboratory: http://www.jpl.nasa.gov/
Lunar and Planetary Institute: http://www.lpi.usra.edu/
National Aeronautics and Space Administration: http://www.hq.nasa.gov/
Solar Data Analysis Center: http://umbra.gsfc.nasa.gov/
Space Telescope Science Institute: http://www.stsci.edu/public.html